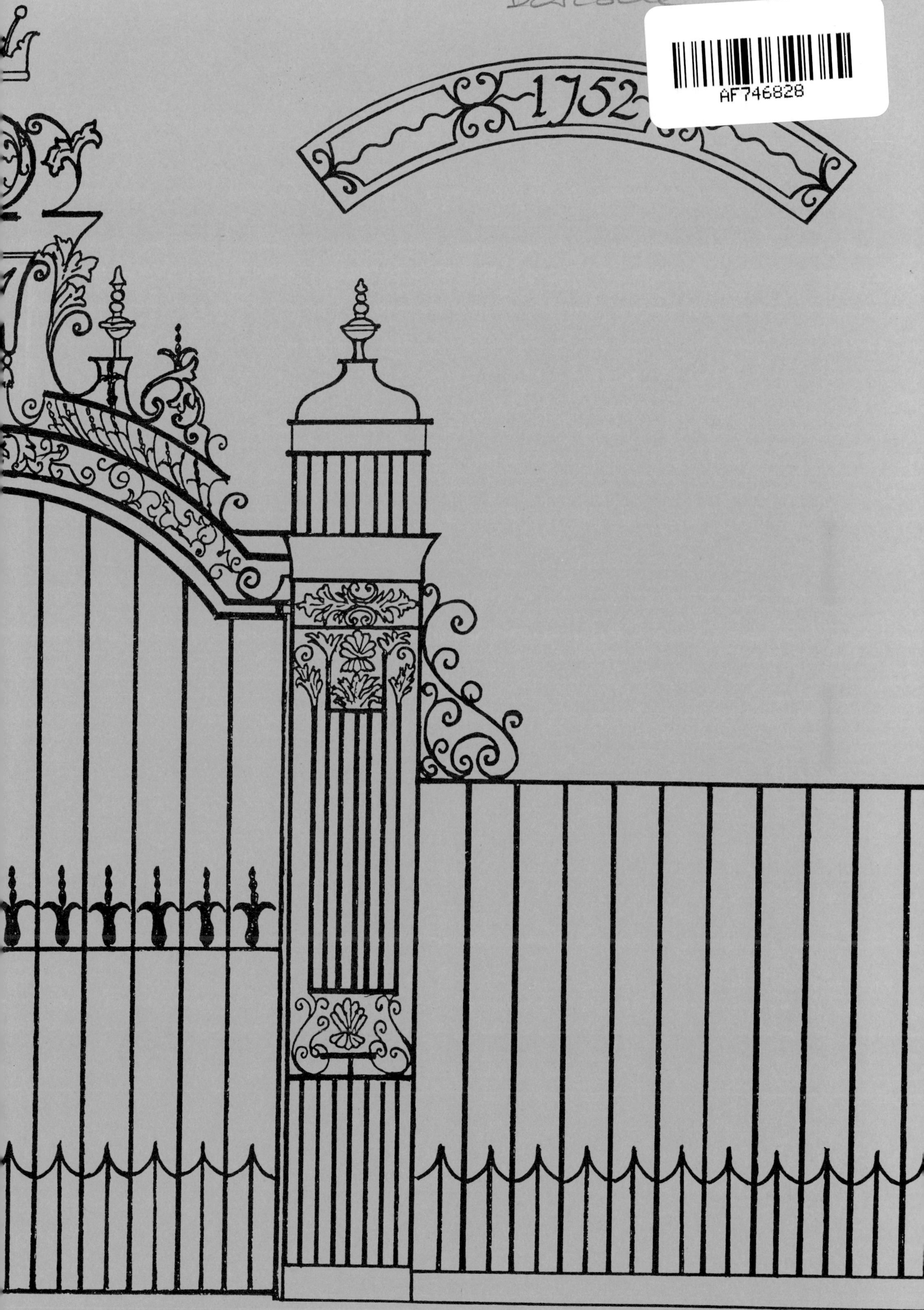
1752

This edition is limited to
750 copies
of which this is No. 176
S. Dunkerley

Robert Bakewell   Artist Blacksmith

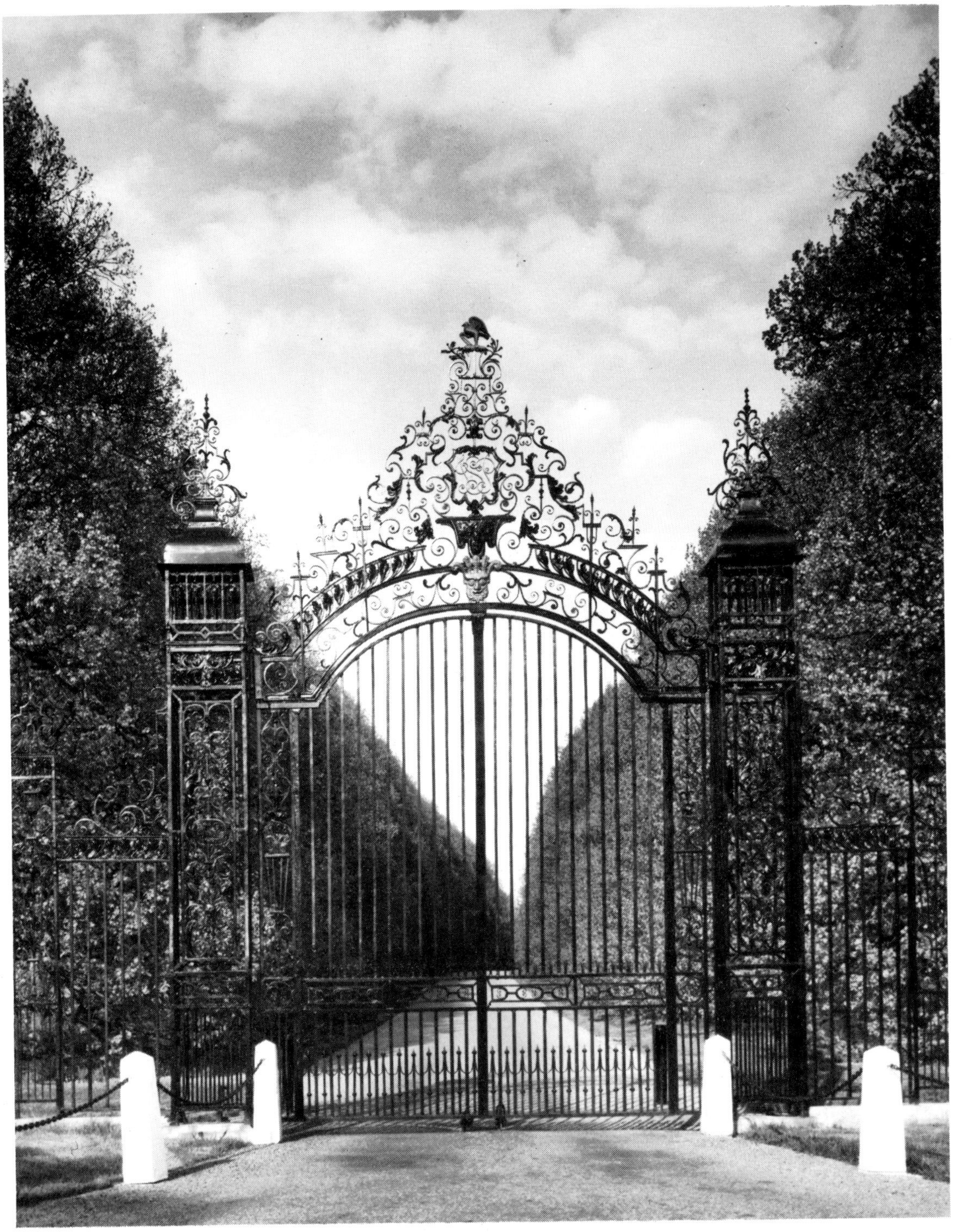

The Gates at Westbury Gardens, New York.
Formerly at Combermere Abbey. *(See page 30).*

ROBERT BAKEWELL

Artist · Blacksmith

S. Dunkerley

Scarthin Books Cromford 1988

Published by Scarthin Books, Cromford, Derbyshire

Phototypesetting, printing by Higham Press Ltd., Shirland, Derbyshire

ISBN 0 907758 24 X

The publishers thank the following and other, unlisted subscribers for their confidence in, and support of, this project.

Architectural Association Library
J. L. Bakewell, *Belper*
Mrs. C. I. Beattie, *Dalbury*
J. Blackwood, *Winshill*
Boston Athenaeum, *U.S.A.*
Mrs. A. P. Bowmer, *Bole Hill*
British Artist Blacksmiths Association
Chatsworth House Library
Chesterfield College of Technology and Arts
Clark Art Institute Library, *Williamstown, U.S.A.*
Clulows Bookshop, *Derby*
Dr. P. H. Cooper, *Etwall*
Professor A. Cox, *Chester*
Maxwell A. J. B. Craven, *Derby*
Rt. Hon. Viscount De L'Isle, *Penshurst Place*
Derby Industrial Museum
Derby Museum and Art Gallery
Derbyshire College Library
Derbyshire Library Service
Derbyshire Museum Service
H. G. DeVille CBE, *Sonning on Thames*
A. Dunkerley, *Glossop*
J. B. Evans, *Winshill*
Mrs. R. Evans, *Alvaston*
Peter Fisher, *Brighton*
Peter Giffard, *Chillington Hall*
P. H. Haslam, *Allestree*
John Heath, *Castle Donington*
J. B. Henderson, *Etwall*
P. P. Housley, *Crowdecote*
P. Johnson, *Ratho, Midlothian*
Mrs. M. Jones, *Llanrhydd Hall*
A. F. Keeble, *Little Dunham*
Margaret Kempster, *Ballyvartry, Ireland*
Lord Ralph Kerr, *Melbourne Hall*
E. J. Knobbs, *Park Hall, Leigh*
T. J. Larimore, *Heanor Gate School*
Mrs. L. K. E. Leahy, *Cork*
Colin Lee, *Lower Hartshay*
Leicester University Bookshop
Dr. G. G. Lennox, *Long Eaton*
Derek Limer, *Derby*
Michael Malleson, *Trent Smithy, Sherborne*
K. A. Miller, *Langford, Essex*
Mrs. R. Milward, *Barlow Woodseats Hall*
J. J. Morrissey, *Belper*
D. W. H. Neilson, *Catton Hall*
University of Nottingham Library
P. J. Oberon, *Linthorpe*
S. R. Penny, *Sheffield*
R. L. Platts, *Mickleover*
Dr. & Mrs. C. J. Pountain, *Cambridge*
Miss F. A. Power, *Sheffield*
Miss S. L. Power, *Belper*
Repton School Library
E. S. Robbins, *Sheffield*
Mrs. D. Robinson, *Littleover*
Marie Robson, *Wirksworth*
Mrs. C. Salisbury, *Burton*
J. F. H. Salisbury, *Willington*
Mrs. M. Salisbury, *Mayfield*
Sheffield City Polytechnic Library
Jonathan Skipper, *Surlingham Forge*
Garry Smith, *Bakewell*
Staffordshire County Library, *Burton*
Derek Stephenson, *Blackfordby*
M. D. Stephenson, *Breadsall*
Leslie Sutcliffe, *Alfreton*
John Thurman, *Solihull*
Gladwyn Turbutt, *London*
Peter Walker, *Chapel Cottage Forge, Anslow*
Walters Art Gallery Library, *Baltimore, U.S.A.*
J. Warburton, *Littleover*
A. B. Williamson, *Kilbourne*
Adrian Woodhouse, *Snitterton Hall*
R. Yeomans, *Stretton*

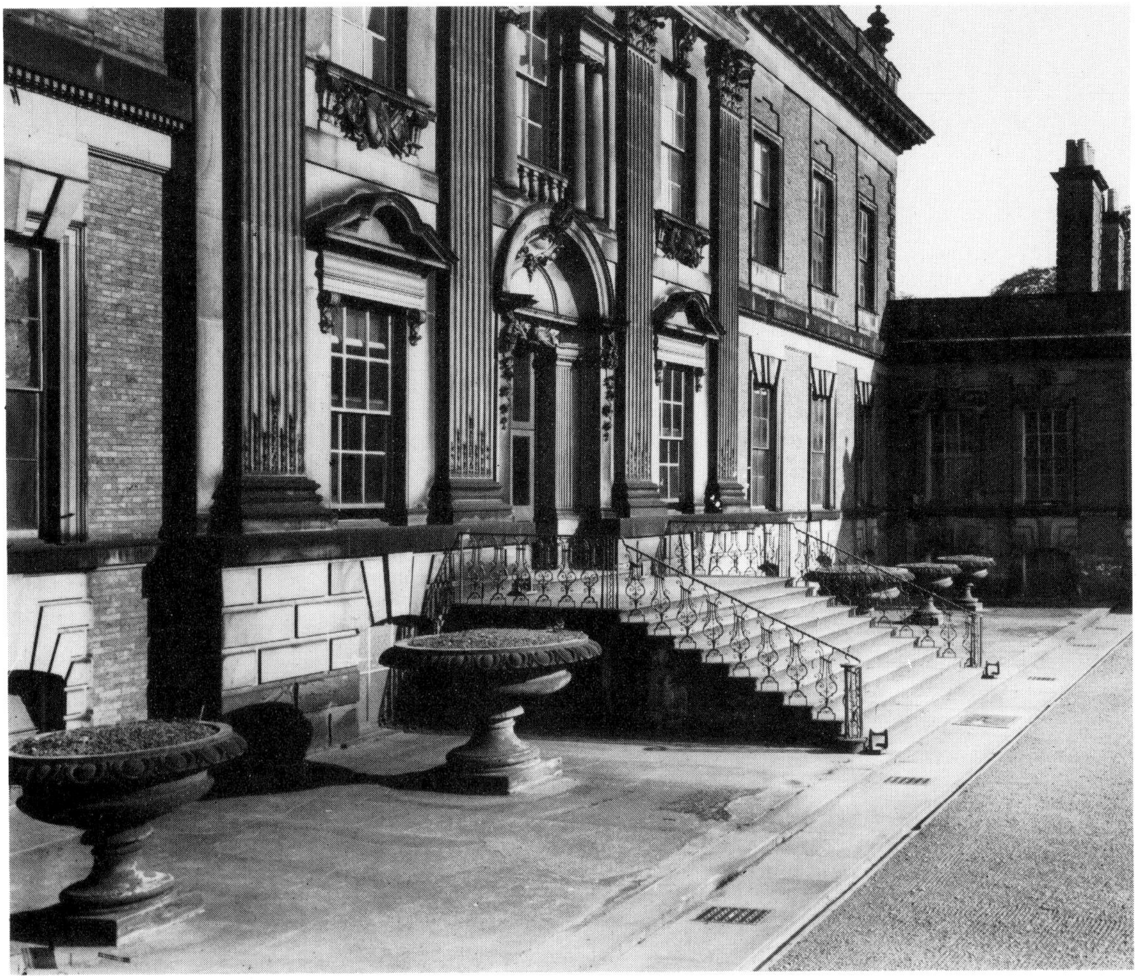

Wentworth Woodhouse. The West Front.
 (See page 66).

CONTENTS

THE FIGURES

THE PLATES

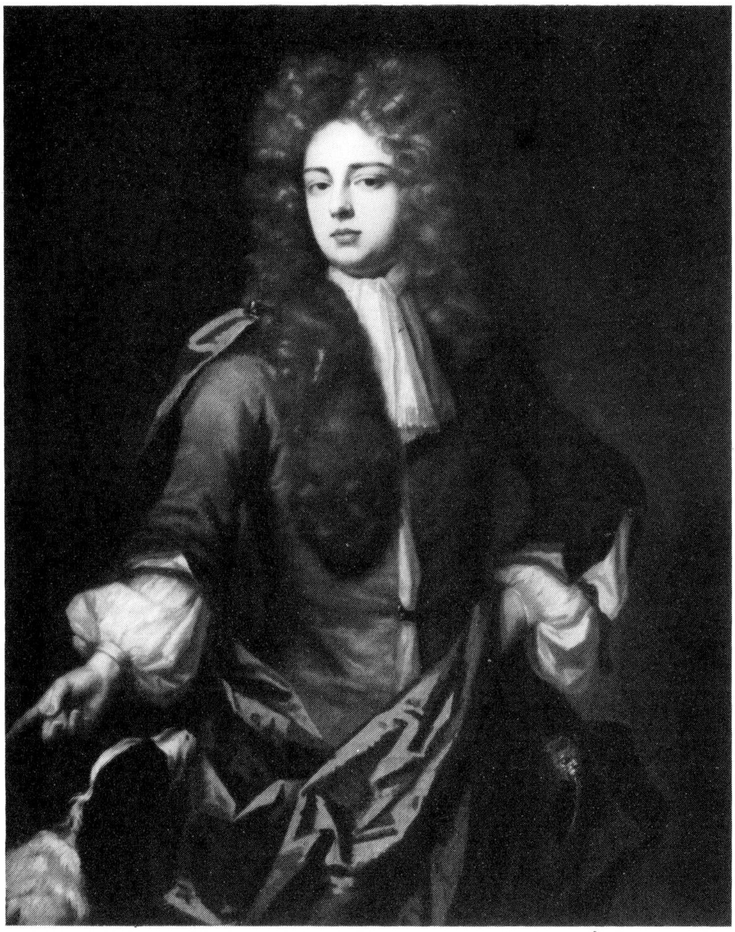

Thomas Coke. Bakewell's first patron.

ROBERT BAKEWELL, 1682 - 1752

INTRODUCTION

Robert Bakewell was baptised at the Parish Church of St Mary, Uttoxeter, Staffordshire, on March 5th, 1682, the son of Sampson Bakewell and his wife Mary (née Sales), whom he married at Saint Alkmund's, Derby. Sampson Bakewell was a blacksmith and Mary was the daughter of a Derby blacksmith. She was only nineteen years of age when Robert was born, and later she gave birth to a son, Thomas and two daughters. Their married life together was short-lived as Sampson died in 1694 leaving two harpsichords, a cabinet of viols, a goodly number of books and the impression that he was a man of some substance. Later (April 28th, 1696), Mary married Lewis Shutt who was probably an employee, possibly the foreman of Sampson, in order, presumably, to save the business for the family.

In 1696 when Robert was fourteen years of age it is believed that he was apprenticed to one of a group of blacksmiths working in London. Two facts emerge to bolster this belief. Matthew Bakewell, Robert's uncle was a carter who plied between the Uttoxeter area and London, frequently carrying craftsmen and their equipment and putting the city within the reach of his nephew. In addition we know beyond possible doubt that Bakewell made railings for a house in Saint James's Place, London, for Thomas Coke. There is documentary evidence to support this amongst the muniments of Melbourne Hall, Derbyshire which was at that time the country home of Coke as distinct from his city house in Saint James's. The agreement concerned with the railings reads:

> June 27th, 1706, Ironwork done for ye Rt. Honr. Thomas Coke Esq. Agreed for Iron Rails at ye House in St. James Place, London, Weight 6 cwt 2 qrs 6lbs at 6d per pound £18 7s 3d.

Thomas Coke was a keen amateur architect and garden designer who kept his position as Vice-Chancellor to both Queen Anne and George I mainly by virtue of his prowess in amateur pursuits, as, apparently, people at Court valued his advice. His home at Melbourne was in the capable hands of his sister, Elizabeth, whose letters help to shed light on Bakewell and his work.

There was a spate of house-building in the late seventeenth and eighteenth centuries so there was a growing demand for sound craftsmanship. The rich and the nouveau-riche vied for the best work and Bakewell, by an accident of time, concluded his apprenticeship when he could achieve maximum benefit from his unbounded skill and his industry.

All craftsmen required a sponsor and in the case of Bakewell the patronage of Thomas Coke helped him on the road to success.

In 1688 King William III and his wife, Mary, daughter of the Stuart King James II came to the throne. They were quickly followed from Holland to this country by a clever Huguenot craftsman by the name of Jean Tijou who was an expert in the craft of Repoussé, in which three dimensional forms are beaten out of sheet metal from the back of the sheet and then sharpened by chasing from the front. It is doubtful if he was a blacksmith as such but he did require good blacksmithery to which he could attach his own work. There was a group of craftsmen who filled this need in London, and it is believed that Robert Bakewell was apprenticed to one of these.

Tijou carried out a great deal of work for William and Mary at Hampton Court, most of which is still standing, some of it rearranged. This shows how florid and complex was his work when left entirely to him. He also worked at Saint Paul's in London, where Sir Christopher Wren requested that he toned down or restrained his decorative talents. On the other hand a balcony fashioned by Tijou at Chatsworth House is quite beautiful.

Tijou left England in 1712 and his wife remained for a while in order to collect money owing to him. He left behind his own ideas or interpretations of the Baroque influence together with a book of 'Desines' which was printed in 1693. These ideas set a trend in this country but were simplified to a great degree into an amalgam which labelled itself English - less fussy, less vulgar, some might say puritanical.

Robert Bakewell was not a repoussé expert. He used the process but limited it in the main to his beautiful swirling acanthus (Plate 17) and his hart's tongue fern or water leaves (Plate 23). These he fashioned on the anvil using a long peined hammer. This process built in a sense of movement in the finished work. He could not, however, match the skill and experience of Tijou in producing masks, and used them presumably only when pressed to do so.

The material from which the blacksmiths of the Baroque period fashioned their work was wrought iron. It was hammered out of a bloom which consisted of the metal and its accompanying roughage or dirt which was removed by the hammering process. The beauty of the material lay in the fact that it usually contained 0.1% or less of carbon. Carbon plus iron in varying greater proportions rendered the material crystalline in structure, but the minimum content of 0.1% or less gave the material a long fibrous nature which, at the correct temperatures, flowed beautifully under the hammer of a bright blacksmith such as Bakewell.

Where Tijou would literally decorate everything and so produce a screen in the fullest sense of the word, Bakewell used his innate sense of proportion as a basis of design and then generally limited his decoration to certain areas. When fashioning gates, for instance, he designed them to stand tall, uncluttered and in sound proportion, whilst his decoration was usually limited to his overthrow, pilasters and when used, his side panels. (See Figs. 1 and 2).

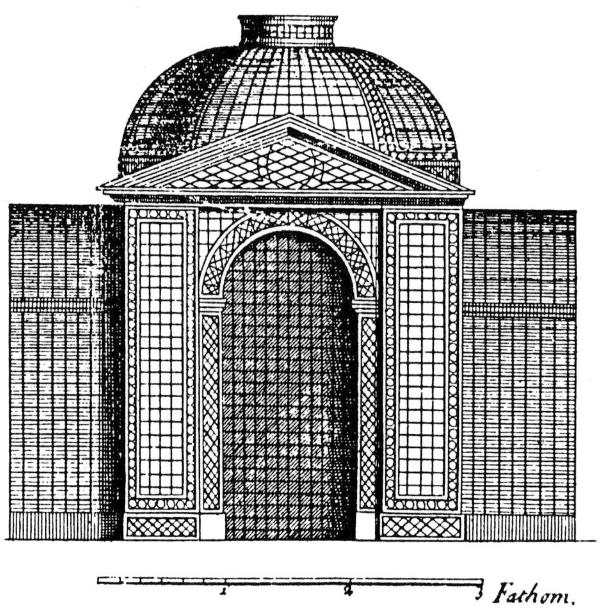

A salon for ye entrance of an arbor. By permission, British Library.

Bakewell's work in London obviously gave Thomas Coke a certain amount of satisfaction and as a result of this he gave Bakewell the commission for the construction of a garden arbour. The French had a fashion for wooden trellis-work arbours to adorn their gardens. Examples can be seen in the painting by Nicholas Lancret below and a picture taken from a book *Le Theorie et la Pratique de Jardinage* published in Paris in 1709 (Page 14).

Coke commissioned Bakewell to build one, not in wood but in wrought iron, not in London but at the family home of Melbourne Hall where he and others he had engaged were busy laying out the beautiful gardens.

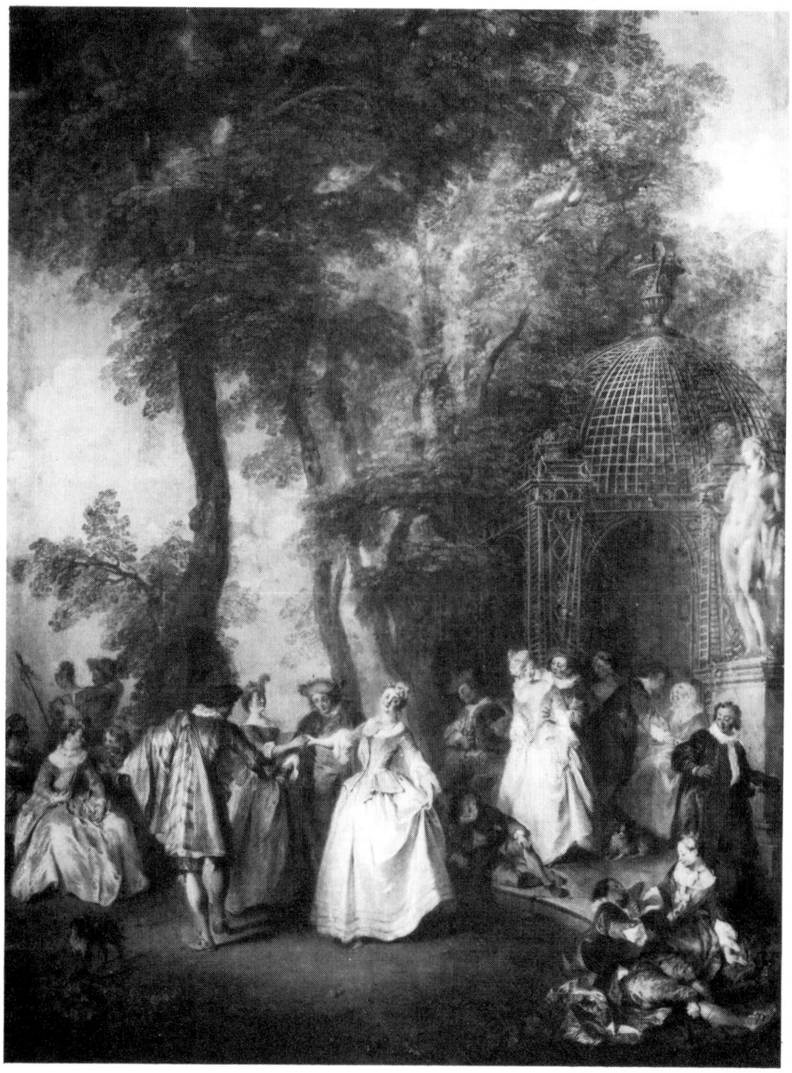

Le Moulinet by Nicholas Lancret, showing a wooden trellis arbour in a french garden.

In order to facilitate the work for Melbourne Hall Bakewell applied through the agent, John Sargeant, to use the forge which was situated in the basement of the Stone House which stands in the Churchyard area opposite the Hall. Plate 1, Top right shows the basement storey, where stood the forge as seen from the back garden. Bottom right shows the house as viewed from the Churchyard, a storey higher, and the plaque which commemorates the 300th anniversary of Bakewell's birth. This plaque was given by E. J. Saunders R.I.B.A., and unveiled by Lord Lothian. It can be seen more clearly at the Centre right of Plate 1. Centre left is a view of the forge, discovered some years ago when a brick wall was removed. Bottom left is Bakewell's quenching trough, where the hot iron was cooled when required.

It must be emphasised that the Stone House is still a private residence and as such, is not open to visitors.

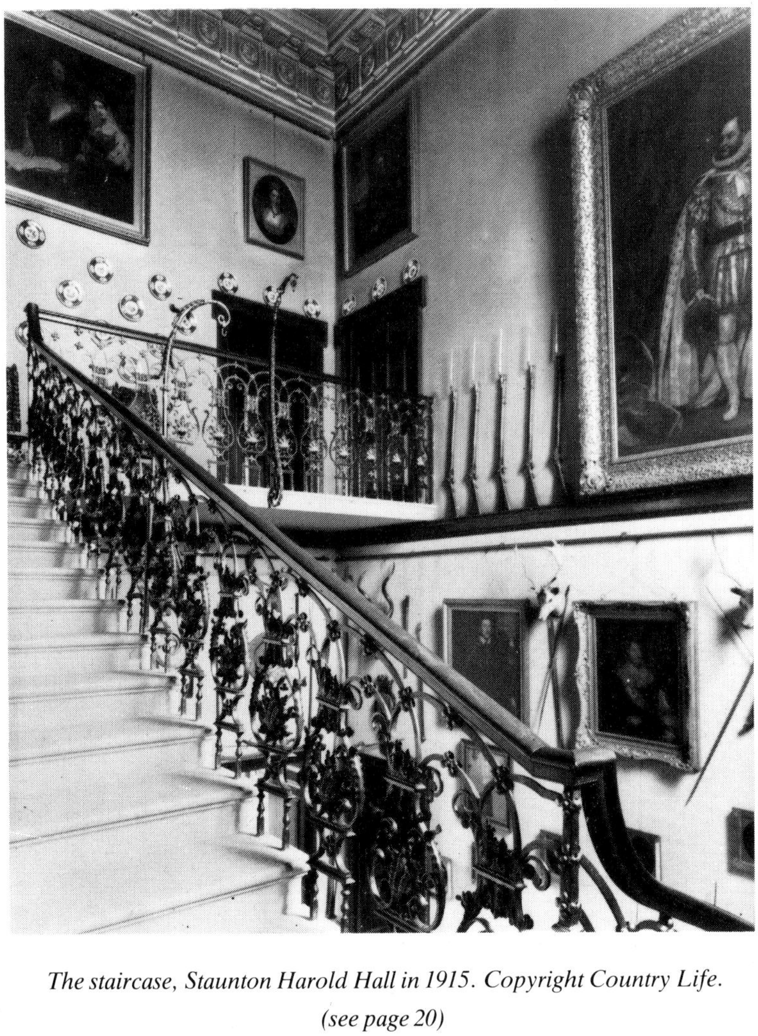

The staircase, Staunton Harold Hall in 1915. Copyright Country Life.

(see page 20)

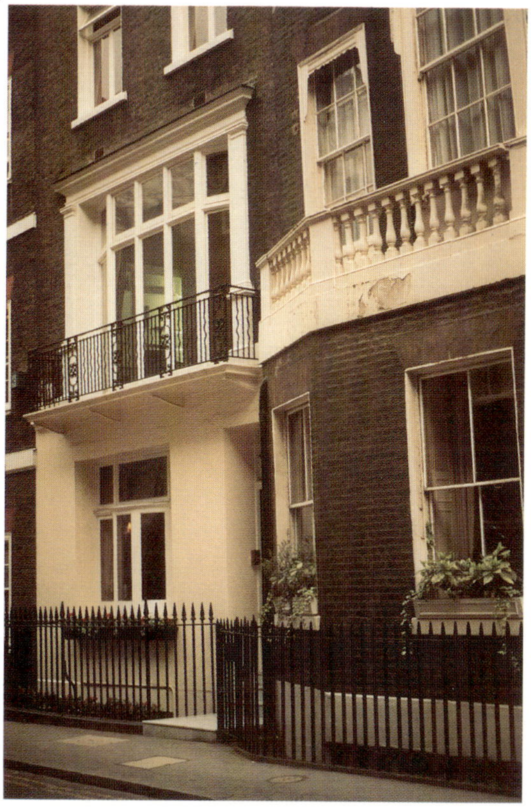

St. James Place, London.

Melbourne, The Churchyard.
The Stone House & Smithy.

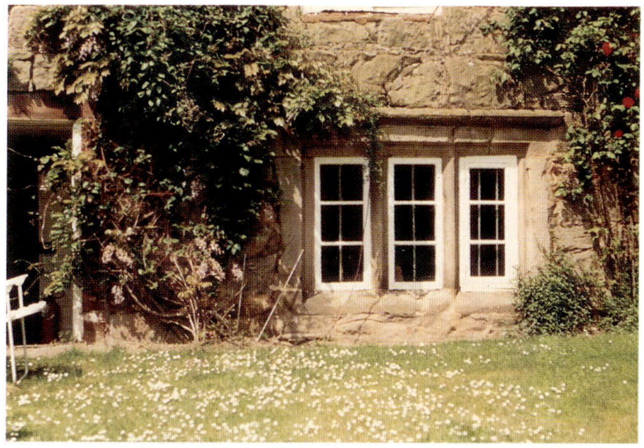

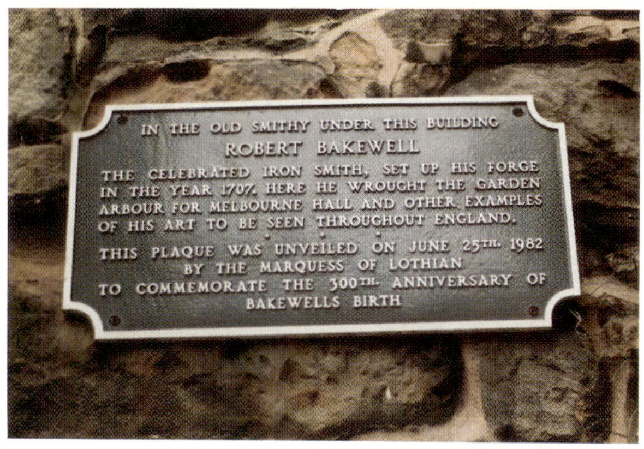

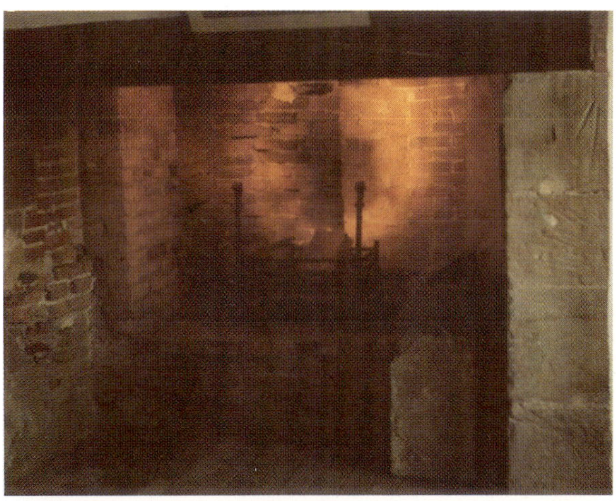

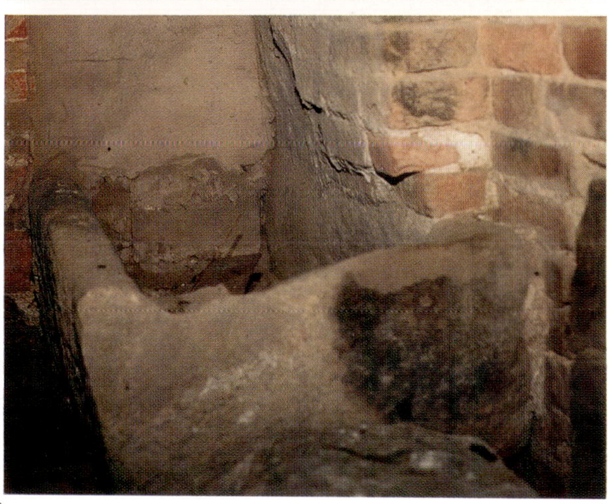

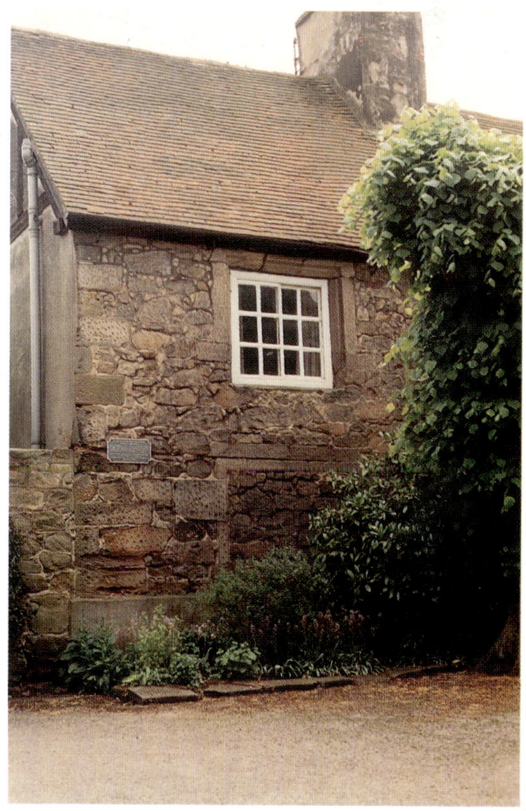

ROBERT BAKEWELL AT MELBOURNE

Bakewell started the making of the arbour in 1707 and probably took two years to complete the initial pattern. We know from one of Elizabeth Coke's letters to her brother Thomas that he did have a long illness during the earlier part of his stay at Melbourne, delaying the completion of the project. (Elizabeth dwelt at the Hall and her letters are a valuable source of documentary evidence). Bakewell was working on artefacts other than the arbour. In one of her letters she refers to work for my Lord Gore (Gower of Trentham), Lord Chesterfield (of Bretby) and my Lord Ferrers (of Staunton Harold). In 1711 the cupola was added to the arbour, as was a small panel at the back where the dome meets the brickwork. For these two items he was paid £6.0.0. His total payment for this magnificent piece of work was £126.0.0. Plate 2 shows that the arbour is actually a decorative addition to an open stone and brick summer house. Standing in the centre of the porch and looking skywards one can see as silhouettes, sixteen intricate panels, eight of which are filled by delicate, cursive scrollwork. Interspersed are the remaining eight, consisting of branches bearing leaves of oak alternating with laurel. Fronting the huge dome is a certain amount of repoussé which has been the subject of some criticism, particularly the mask which has been described as comic, ugly and poor compared with Tijou's masks. However, the mask apart, there can be little other valid criticism. No one has ever denigrated Bakewell's ironwork and his acanthus and water leaves leave nothing to be desired. The arbour is visible from the Hall across sweeping lawns and a fair sized lake, adding its own charm to the pastoral scene.

The writer is not alone in his praise of the arbour. For instance Betjeman, in his *Guide to Parish Churches (1980)* writes "Immediately to the East of the Church (Melbourne) is the Hall, with a fine garden featuring a great basin and an exquisite wrought iron Arbour by Bakewell of Derby 1706 - 1711".

Gary Hogg in his *Guide to English Country Houses* also refers to Melbourne Hall and gardens and concludes his passage as follows: "But it is the Birdcage, a masterpiece in wrought ironwork by a Derbyshire blacksmith that draws and holds the eye before and after all else; a miracle of craftsmanship in formal-natural setting of velvet lawns, flower beds and the hint of the lake beyond". Perhaps Raymond Lister in his book *The Craftsman in Metal* pays the most appropriate tribute. He refers to "A craftsman like Robert Bakewell who made the domed garden arbour at Melbourne near Derby, so delicate that it would not be out of place as a setting for the last act of Mozart's Nozze di Figaro". It is appropriate that the birdcage or arbour is now a Grade 1 listed building.

During his stay at Melbourne Bakewell also constructed the attractive garden balustrade which overlooks the lake. The lyre form seen in the balusters was frequently used in Baroque ironwork and particularly by Bakewell. The shell form can also be seen as can the spur shown clearly in Fig. 3.

BAKEWELL FISHER

Not all Bakewell's interest during his stay at Melbourne was focused on wrought ironwork. In the Parish Registers of the Church of Saint Mary and Saint Michael, Melbourne, dated 1709 October ye 26 appeared the following line 'Bakewell ye basterd child son of Robert Bakewell and Elizabeth Fisher was baptised'. The Fisher family, mother and five daughters, lived in the Stone House above the forge where Bakewell worked. This incident and possibly others, soured the relationship between the blacksmith, the Cokes and the Fishers. He still worked on in the same forge, however, and produced much delightful work there until he left in 1711 to open a smithy in Oakes Yard in Derby.

PLATE TWO

Melbourne Hall, Derbyshire,
The Bakewell Masterpiece —
The Wrought-iron Arbour —

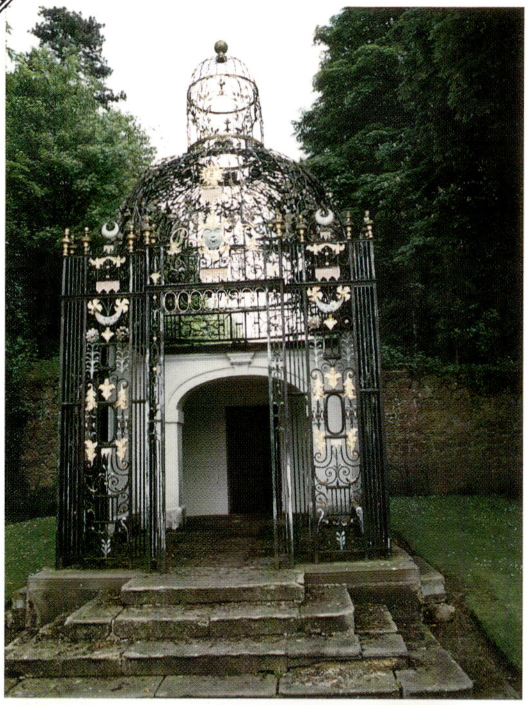

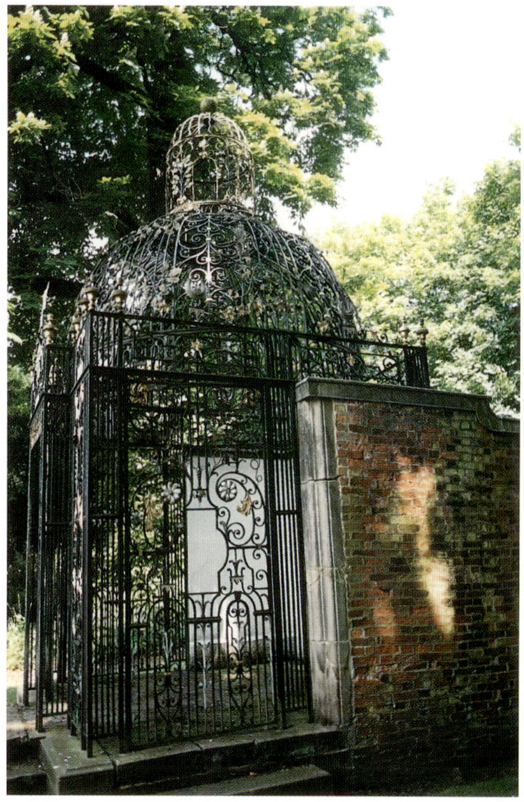

Now a 'Grade One' Listed Building

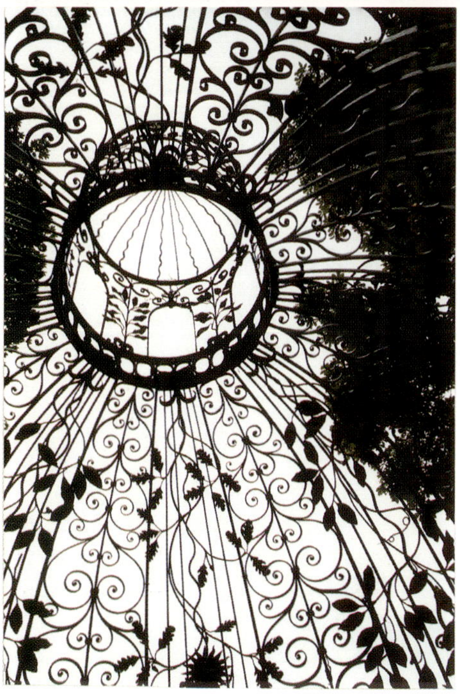

and Garden Balustrade

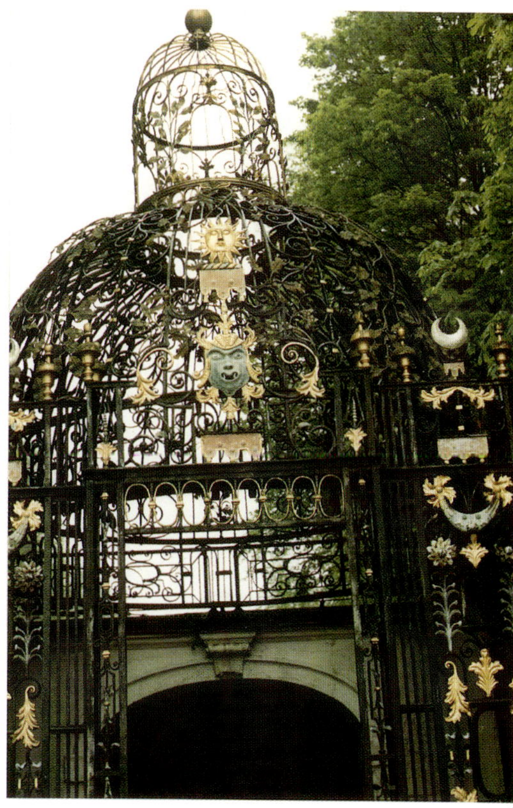

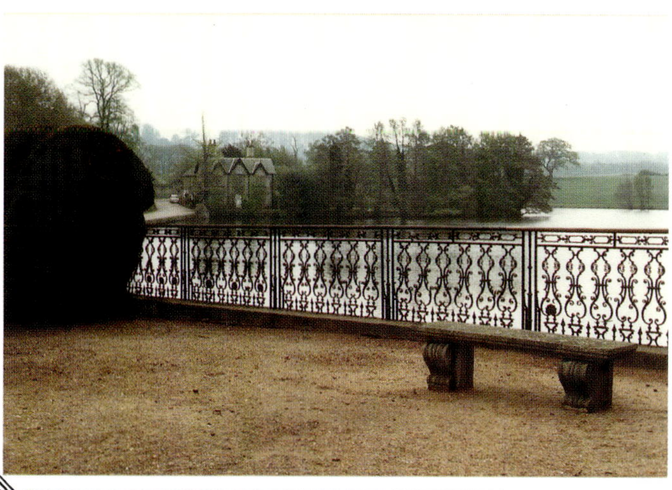

BAKEWELL AT STAUNTON HAROLD

Staunton Harold is best known for its Hall, now the headquarters of the Ryder-Cheshire Mission. Formerly it was the home of the Shirley family, later the Lords Ferrers. Robert Bakewell's contribution to the Hall was the magnificent staircase with its linked balusters, a pattern which he also used at Okeover Hall and the Maister's House, Hull. (Fig 3). The staircase is shown on page 16.

The exact date of the making of the staircase is not known but it must have been early in the eighteenth century. The first house to be built at Staunton was erected in 1324 by Sir William de Staunton. In 1423 Sir Ralph Shirley married Margaret de Staunton bringing the Hall into the Shirley Family. The Palladian style building, as we see it today was built by the fifth Earl Ferrers (Washington Shirley) in the eighteenth century.

In 1653 the Church of the Holy Trinity was built close to the Hall by Sir Robert Shirley. This he did in defiance of Cromwell who, because Sir Robert refused to finance a company of soldiers for him, sent him to the Tower where he died at the early age of twenty eight years. Over the West door of the Church one reads the following:

In the year 1653

When all things Sacred were throughout the nation

either demolisht or profaned

Sir Robert Shirley, Barronet,

Founded this Church;

Whose singular praise it is,

to have done the best things in the worst times

and

hoped them in the most calamitous.

The righteous shall be had in everlasting remembrance.

In the Holy Trinity section of the National Trust pamphlet on Staunton Harold appears the following passage: "The Chancel which is relatively low, is separated from the Church by a depressed arch which is spanned by a splendid wrought iron grille, incorporating the Shirley arms. The grille is by the famous smith Robert Bakewell of Derby, with whom Lord Ferrers placed an order for ironwork in 1711. It may be compared with Bakewell's Screen in Derby Cathedral which dates from 1725 and with his work at Foremark Church and at Melbourne nearby".

As an exercise in sound design and craftsmanship the screen leaves nothing to be desired. It fulfilled the role for which it was intended in the eighteenth century but it also left much of the East end visible to those who worshipped there. The lyre motif is dominant and the dog bars (Fig 4) on the wing panels are obvious derivatives of others by Jean Tijou. The scroll work, of which there is an abundance, is of a very high standard.

The woodwork in the Church is an excellent example of Jacobean craftsmanship, and a fitting memorial to William Smith who carried out the work.

PLATE THREE

The Sanctuary Screen Staunton Harold, – the Church of The Holy Trinity.

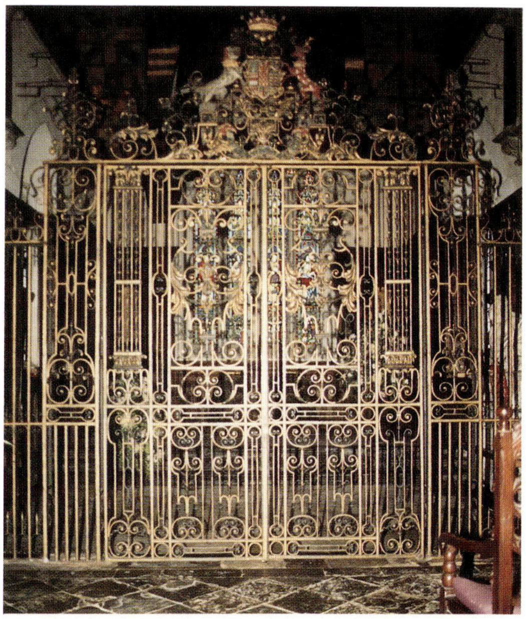

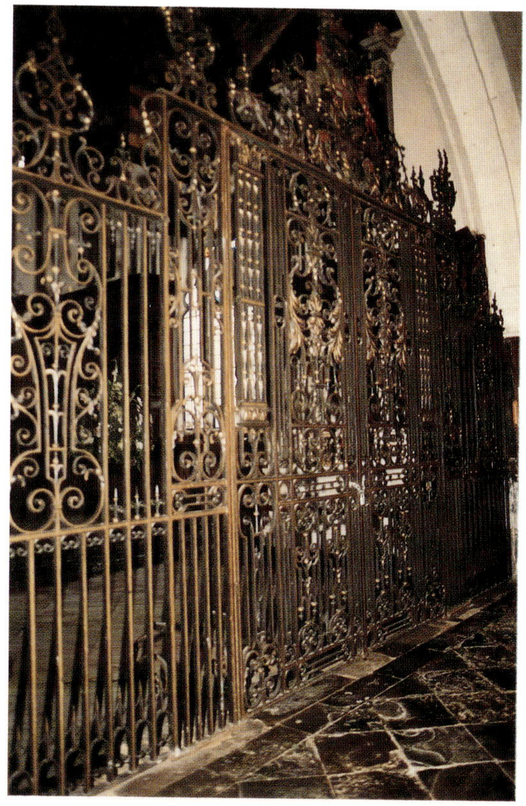

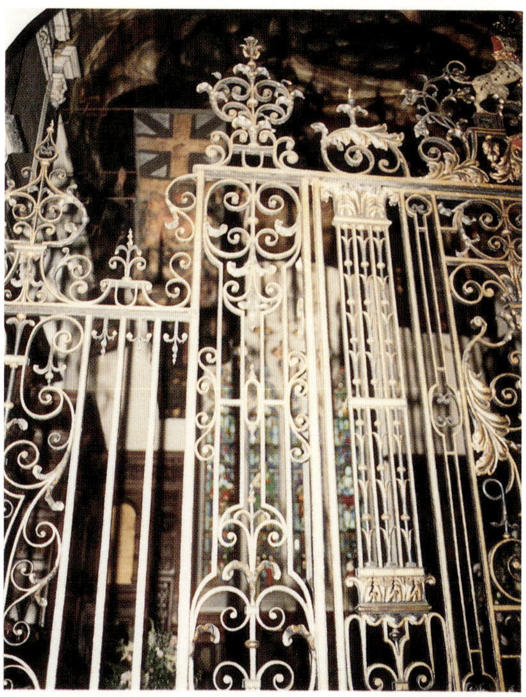

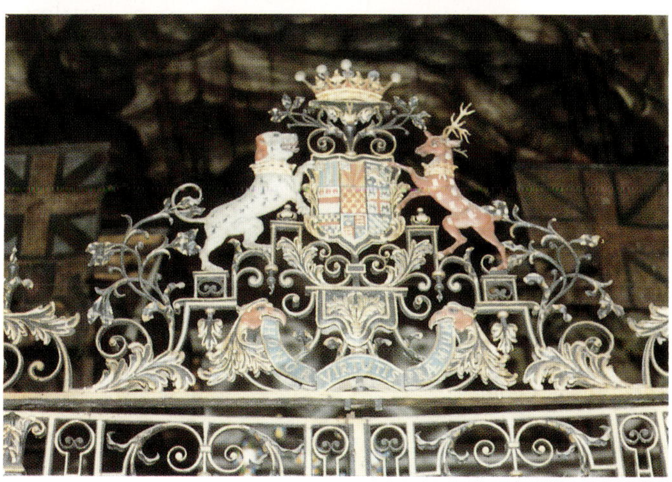

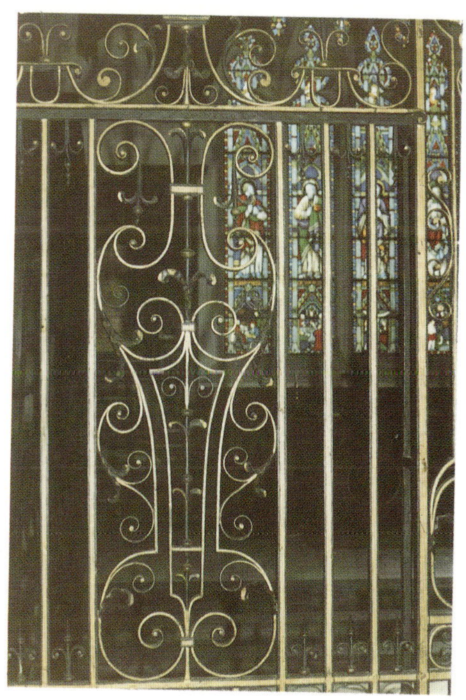

BAKEWELL AT FOREMARK

In the year 1662 the Burdett family built the Church of Saint Saviour's at Foremark. They lived in Foremark Hall, a dwelling they had acquired by marriage earlier in the seventeenth century. The Church is approached by a narrow, rutted cart track off the Repton to Ingleby road and is virtually hidden by a copse. However, once found it repays the effort and the short journey. The original Jacobean woodwork is very similar in character to that of Staunton Harold Church. Though generally rather plain the East end is enlivened considerably by the presence of Communion rails by Robert Bakewell. Excellent work demands and deserves the very best treatment. At Foremark that idea is realised. The rails are finished in light blue and gold, and literally light up the Sanctuary.

Facing West from the Altar are three wide panels separated by couplets of very narrow panels which are of two designs so that each pair consists of both types. On the one hand are thunder and lightning bars (Fig. 4), on the other beautiful scrollwork decorated by acanthus, water leaves and bead work. The centre panel is decorated by a profusion of scroll and acanthus bordering a central cartouche bearing the crest of a lion.

The remaining wide panels have a bar motif in the centre and are decorated by acanthus in both rosette and scroll form and by water leaves.

Gilding of this type of work is a tribute to those who made it. It is an expensive process which is used only where the work appears to merit the treatment. It compares with the use of marble in sculpture; that, too, is reserved for the very best work.

Beyond the East end of Saint Saviour's is an avenue which leads to the Hall. The way is barred by a pair of beautiful but unusually delicate gates, flanked by slender standards (Fig. 5), decorative side panels and sphere-topped stone pillars. Below the lock rail (Fig.1) one finds not dog bars but thunder and lightning bars, as seen in the Communion rails of Saint Saviour's. On the level of the lock bar but in the standards is the inset rectangle with two wavy bars. (Fig. 6). More rectangles with wavy bars are to be seen immediately beneath a joyous overthrow. On each side of the overthrow, resting on the stretcher frame are two repoussé shells (coquillage). At the base of the central open bar finial is a gilded cloth of estate. The wide panels are beautiful crested and topped, as is the overthrow, by a leafy motif. (Fig. 16). All in all the whole screen is light, handsome and dignified.

Late in the year 1711 conditions had so alienated Bakewell and the Cokes that it was obvious that his presence there had become untenable. He was however owed money and found that moving and equipping a new workshop was difficult. His move was eventually expedited by money from Elizabeth Coke, so he left to go to Oake's Yard in Derby, on land owned by the Liversage Charity in close proximity to Saint Peter's Church. This fact was researched by the late Mr J. Austin, a former Headmaster of Mackworth School, a few years ago. The site has now been built over and is no longer visible.

This position must have suited Robert Bakewell as he stayed there for a total of forty years, developing his business and expanding his workforce.

Saint Saviour's Church, Foremark, nr Repton......

Bakewell's Communion Rails and —

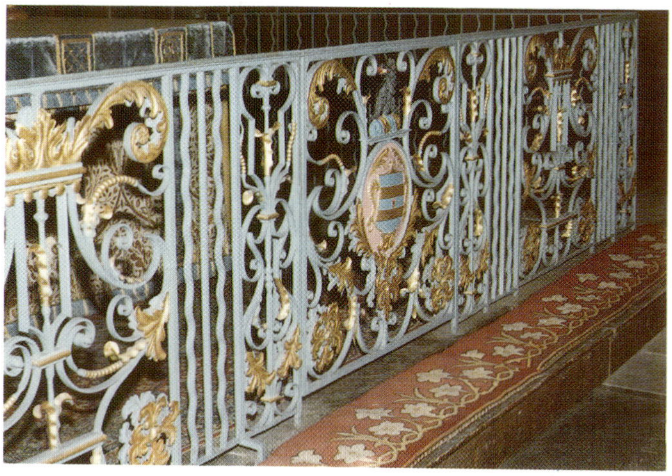

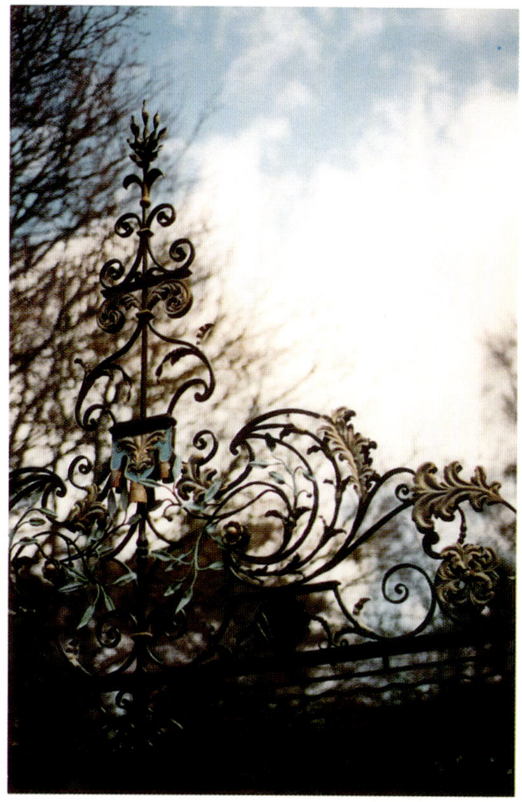

Churchyard or Avenue Gates and Rails.

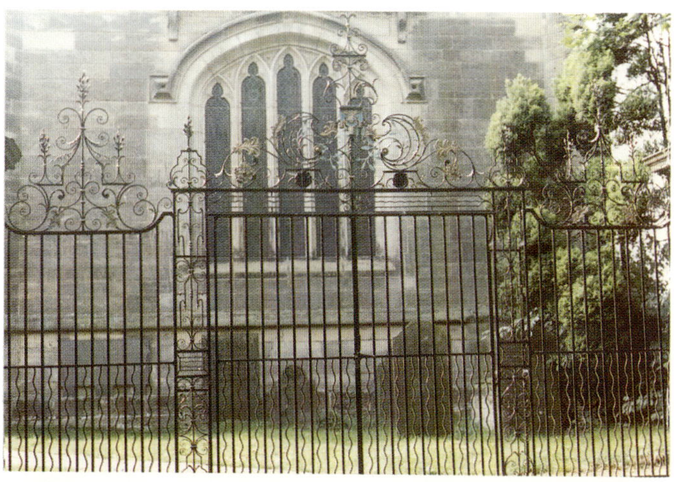

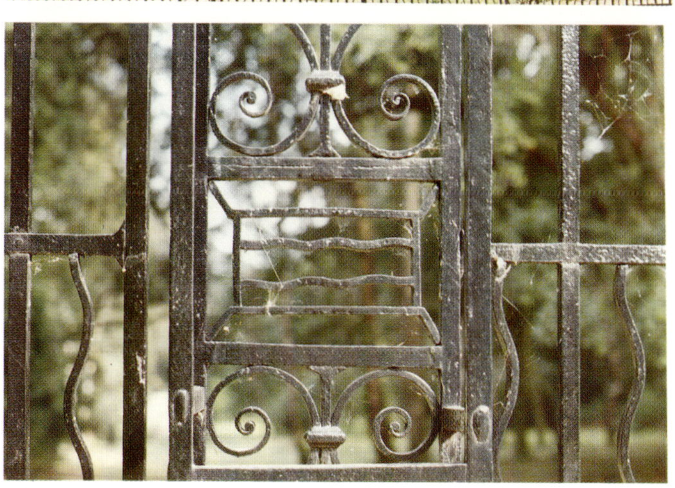

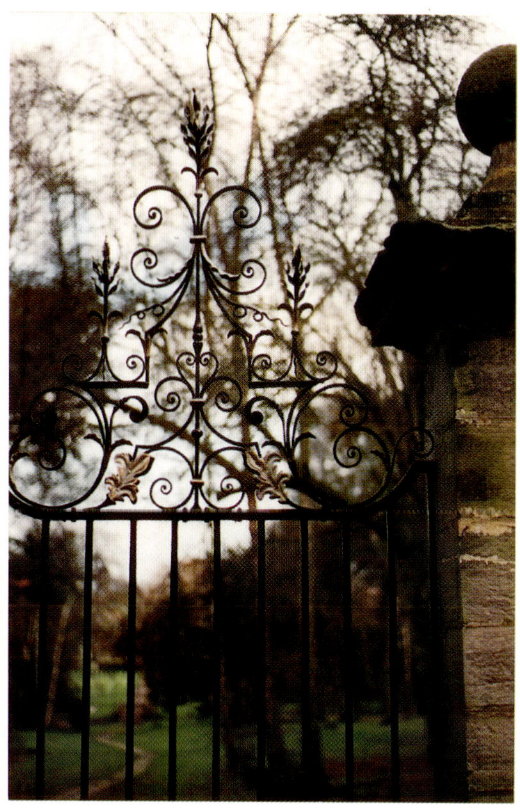

BAKEWELL IN MANCHESTER

Writing in *Country Life* January 23rd, 1975, Edward J. Saunders R.I.B.A. an authority on Baroque building and ironwork, in reference to Bakewell's removal to Derby says, "He had made or was about to make the Communion rail for the Chapel at Foremark as well as its near replica for Saint Ann's in Manchester".

There does not appear to be any direct documentary evidence of Bakewell's involvement in this work; Canon Peter Vowles, however, writes in a letter to the author that the fact the rails are Bakewell's is "well attested."

The Church is probably the best maintained Parish Church that the author has ever seen and the Staff are obviously very proud of the wrought ironwork there.

Although the rails (panels is a more suitable appellation) are not gilded, the swirl of the long scrolls with the accentuating ripples of the water leaves and beads, seems to deny the very existence of the unbending horizontals above and below, filling the eye with unending movement; artistry in iron indeed! Notice too, the inset rectangle in the narrow side panels, and Bakewell's medallion acanthus.

BAKEWELL AT BENINGBROUGH HALL, NEAR YORK

This former home of Lord Chesterfield whose family were once at Bretby near Burton upon Trent, is now in the hands of the National Trust. In the main Hall are four splendid wrought iron grilles of almost identical pattern to the panels forming the Communion rails at Saint Ann's in Manchester. The only difference is the design of the narrow end panels, readily seen in Plate 5, The Pamphlet issued by the Trust suggests that the grilles would be sent by boat along the rivers Trent and Ouse, navigable rivers being the leading form of inland transport at the time. Although the grilles are now in their original position in the Hall, earlier, probably at the beginning of the nineteenth century, a hanging gallery was positioned across the South end of the Hall. This gallery had an unfortunate effect on the proportions of the main Hall and was removed in 1920 thus enabling all four panels to be returned to their original positions. They have, too, been repainted in their original colours.

Plate 5 shows all the grilles as they now are and Top right gives some idea of the height at which they are placed.

BAKEWELL AT CHOLMONDELEY

The story of Bakewell's work at Cholmondeley really starts with the Huguenot Artist Craftsman, Jean Tijou, rather than Bakewell himself. In 1695 Tijou was paid £100 for "Iron Gates at the end of the Garden" and another "little yron gate" was sent along with them by carrier from Tijou's workshop near Hampton Court, where he carried out a great deal of decorative work. This was recorded by Gervase Jackson-Stopps in *Country Life*, July 19 and 26, 1973.

Writing from London to his steward, Mr. Adams, in a letter dated February 14th, 1694/5, Lord Cholmondeley intimated that he had agreed to purchase gates from Tijou and was sending a pattern of two pillars that he wished erected by "Downes the Gardiner". He also stated that they must be 67 feet apart measured from the centre of the base of the pillars.

The very presence of work by Tijou obviously established a precedent and it was natural that when more wrought ironwork was erected there it would be influenced by his design and standards, in order to create some basic uniformity.

PLATE FIVE

The Communion Rails, St. Anne's, Manchester c1712., and the Four Grilles of Beningborough, Yorks...

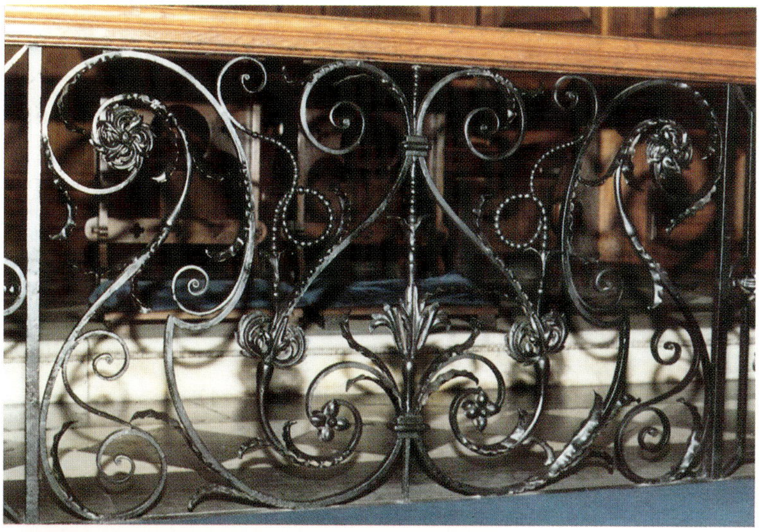

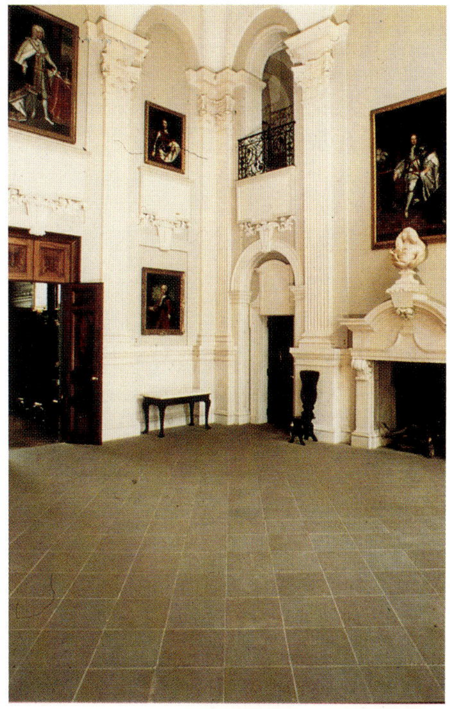

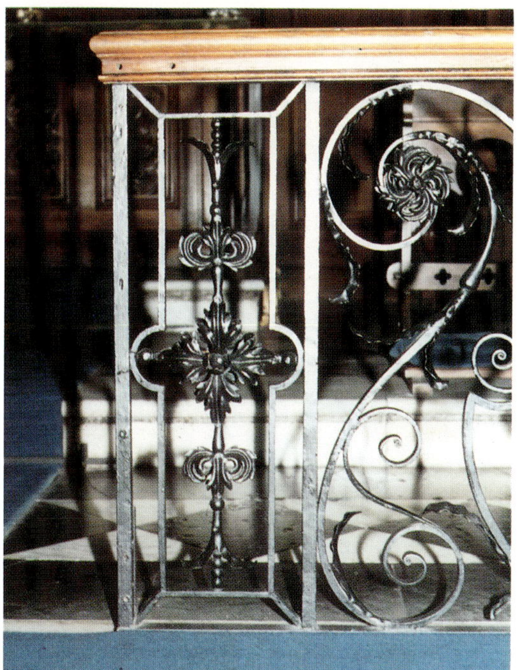

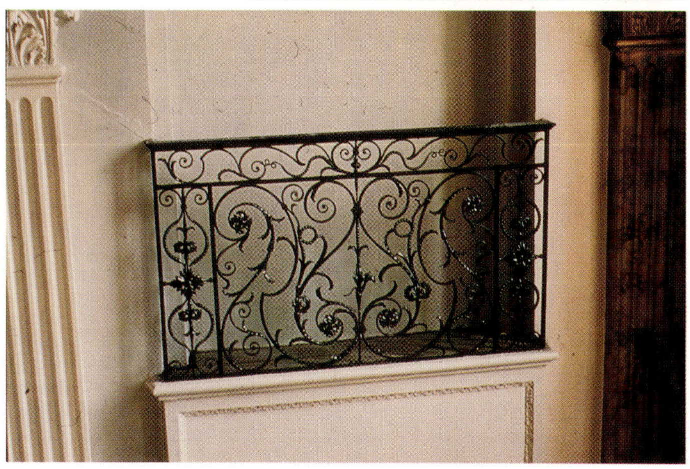

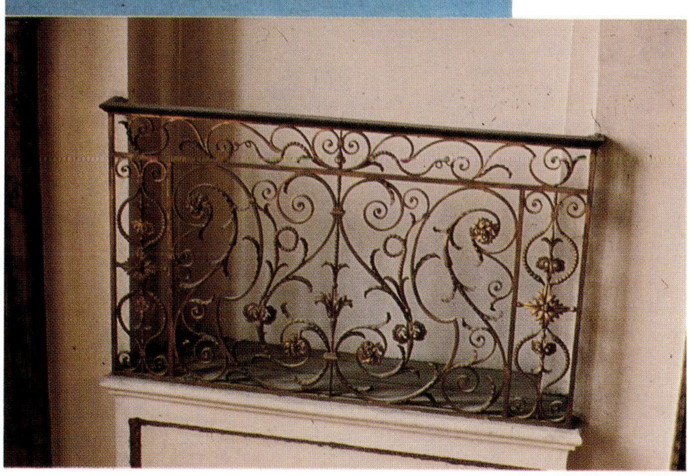

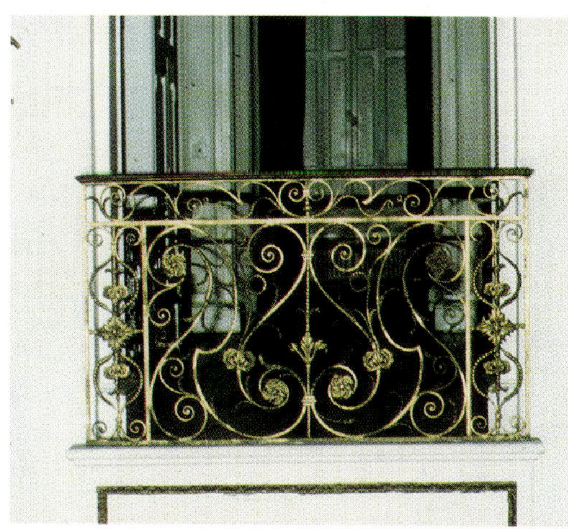

Robert Bakewell, who, by this time was gaining in popularity, mainly it can be assumed by recommendation, drew up an agreement with Lord Cholmondeley for the making of staircase rails and banisters. Transcribed it reads as follows:

> An article of agreement made betwixt ye Rt Honrb ye Earl of Cholmondeley and Rob Bakewell of Melbourn in ye County of Darby For makeing the Iron Rales and banestors for ye Grate Starecase and ye Panels for the Galloreyes about ye Hall is to be made as folloeth viz The aforesaid Rob Bakewell is to make Starecase Rales and benestors according to ye draught annexed viz ye banestors of ye Starecase to be made thirty three inchs Hoigh and eight Pilastors to be Rought Square and Proportionable to ye banestors. The eleven Vaconces in ye Galloreyes is to be ffitd up with a Rought Iron Panol according to ye drought annexed. The Heigh of the eleven panols is to be Three Feett Heigh. For all ye work ye Sd Robert Bakewell is to have one Hundred and Ten pounds to be paid him viz: Fiftey pounds to be Pade him at ye Delivering of the Eleven Panels viz at nantweich ye Whole worke is to be Delivered at ye charge of Rob Bakewell And to be brough from thence to Cholmondeley at his Ld Ships charge. The whole Complement of worke is to be Completed and Delivered by ye next in ye ensuing year. 1711.
>
> The Remr of ye money is to be paid when ye whole is sett up y finisht or within a month of same.
>
> Wm Adams
>
> N W Smith
>
> Robert Bakewell

(Cheshire Record Office DCM/M37)

A further note addressed to "Mr Wm. Adams, Steward to ye Rt Honrb ye Earl of Cholmondeley with a Parcoll May 20 1712", reads as follows:

> Darby May 20 1712. Sr I have sent you 6 Brass lockes to ye Red Loy on in nampt woich with ye mastor kay and two other kays belonging to them from Sr. Your Most Hum Servent
>
> Rob Bakewell
>
> Sr After I had right this note I ordered them to be left at Mr. Taylors.

(Cheshire Record Office DCM/M37)

We are fortunate indeed to be able to quote evidence of Bakewell's work for the Earl of Cholmondeley. Although not everything was dated there is sufficient spread of dates to show that he worked for the Earl from 1711 until 1722. If we return to the top of the page where there is a transcript of an agreement concerning the Staircase and Banisters we will notice that the work was taken to Nantwich at Bakewell's expense but from Nantwich to Cholmondeley at his 'Ld Ships' expense. There is an account slip extant (*Cheshire Record Office DCH/47*) which reads:

> 1713 Nov 11th to Sam Roden for ye carr of thi Iron
> belonging to ye Staircase 01 — 01 — 06.

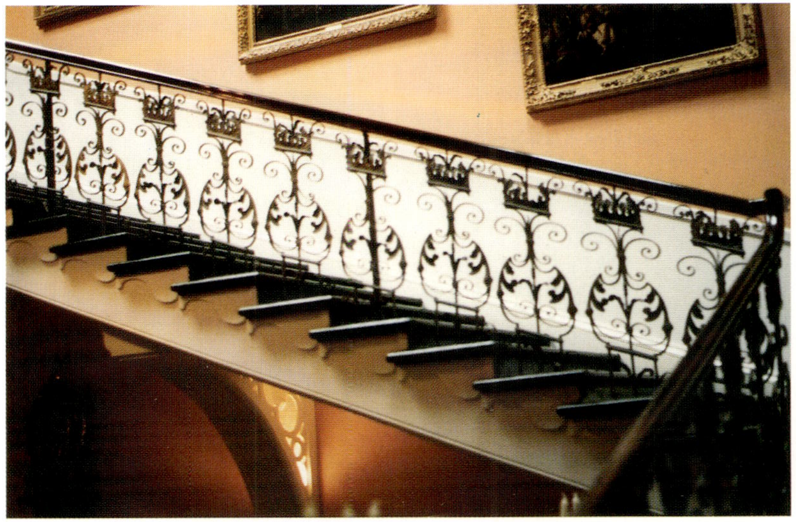

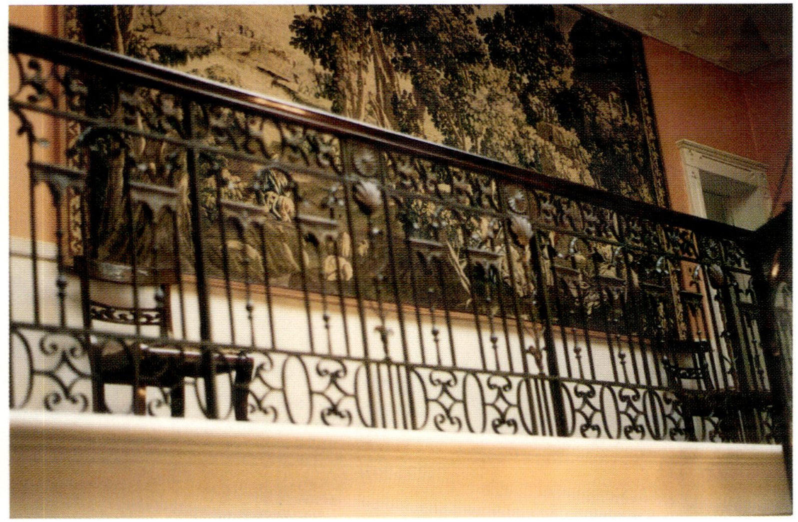

The Staircase and panels mentioned in the agreement of 1711...

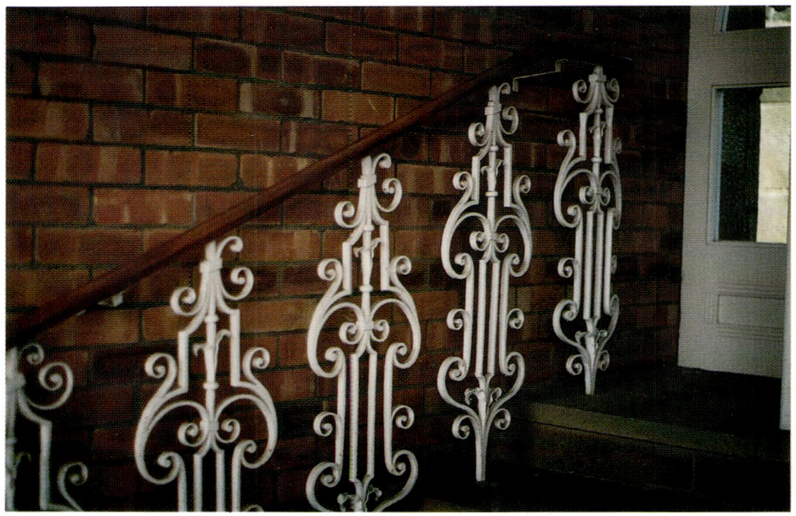

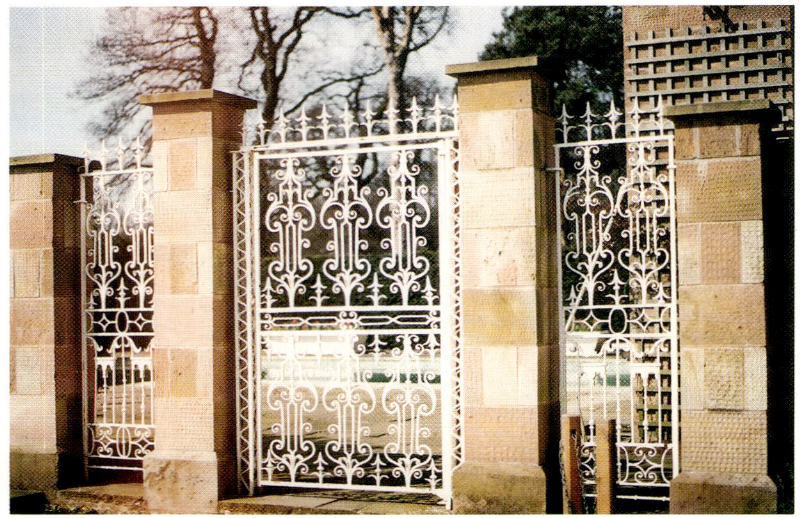

The Loggia with Ironwork from the back stairs.

Cholmondeley Castle, Cheshire.

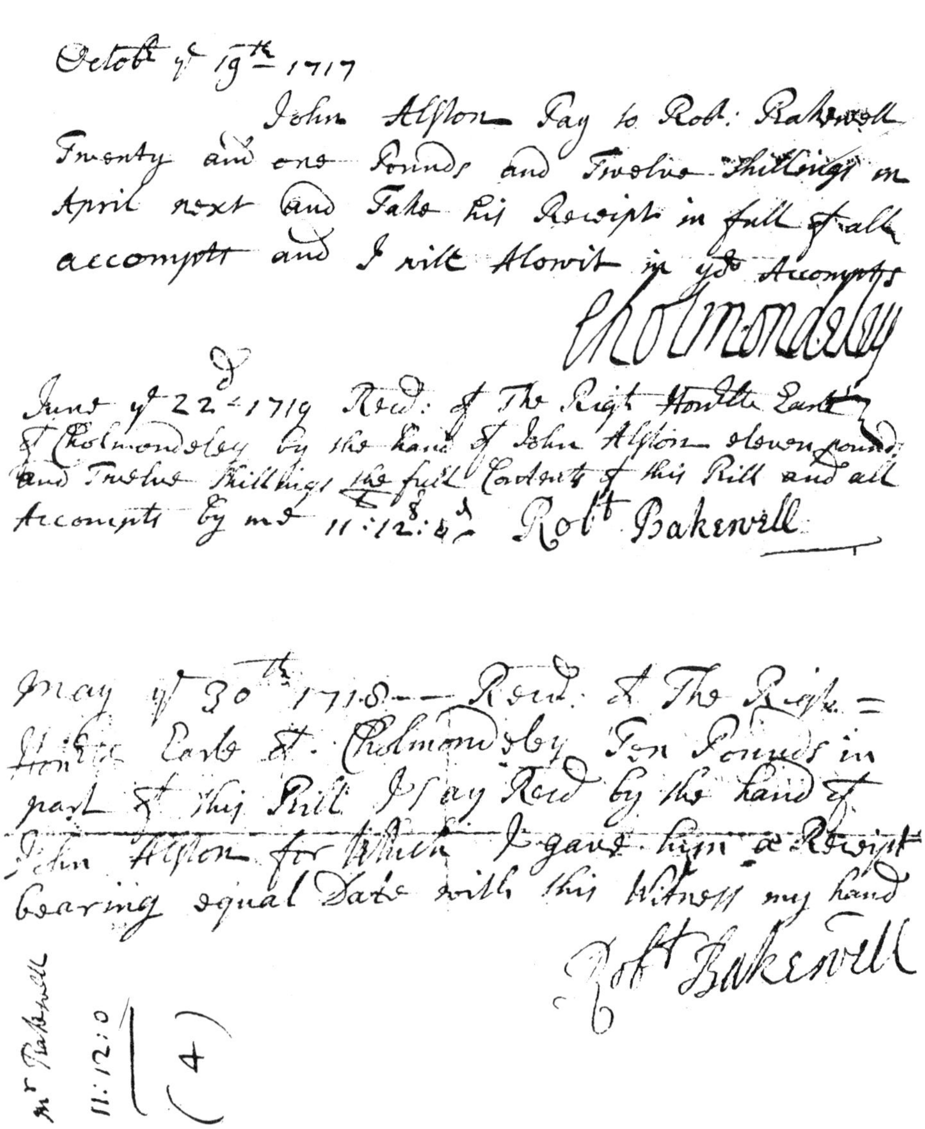
Octobr ye 19th 1717
John Alston Pay to Robt: Bakewell Twenty and one Pounds and Twelve Shillings on April next and Take his Receipt in full of all accompts and I will Allow it in yor Accompts
Cholmondeley

June ye 22d 1719 Recd: of The Right Honble Earl of Cholmondeley by the hand of John Alston eleven pounds and Twelve Shillings the full Contents of this Bill and all Accompts by me 11:12:0 Robt Bakewell

May ye 30th 1718 — Recd: of The Right Honble Earle of Cholmondeley Ten Pounds in part of this Bill I say Recd by the hand of John Alston for which I gave him a Receipt bearing equal Date with this Witness my hand
Robt Bakewell

Mr Bakewell 11:12:0 (4)

Documents dated 1717, 18 & 19.

Bakewell's work for the Earl of Cholmondeley was quite varied. It was made for the house which was started in 1707 and finally completed in 1717. One hundred years later, in 1801 the first Marquess abandoned it and built the Gothic Castle which exists today. This means that all the wrought ironwork has been adapted to entirely new conditions, little remaining where it was originally placed. We now see the White Screen as probably the only outside work in its original position. In addition there is the Chapel Screen and Steps, a balustrade in the Moat garden, Loggia ironwork which is partly adapted from the back stairway and, of course the exceptional stairway and panels in the Main Hall.

A study of the work on the stairway and landing in the Hall will show just how much Bakewell had to adapt to, or was influenced by, work done by Tijou. This work is in contrast to his design at Okeover, Staunton Harold (Page 21) and the Maister's House in Hull. Its atmosphere is continental but fortunately not so crowded with detail as would Tijou's efforts have been. It is by any standard restrained and this is a Bakewell feature in itself.

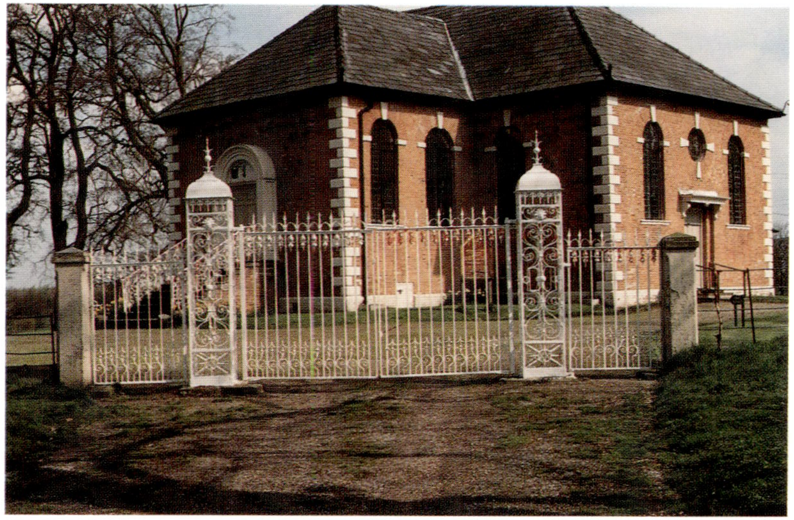

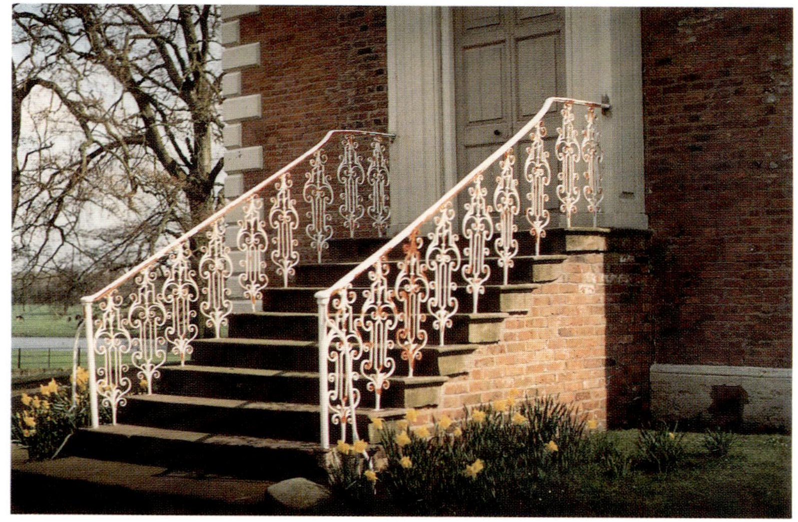

The Chapel Screen and Steps, with details of Dog-bars and fringes.

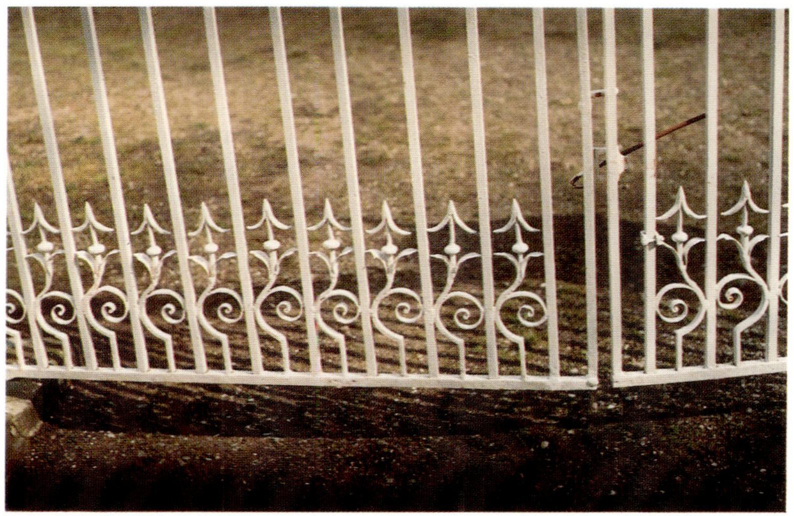

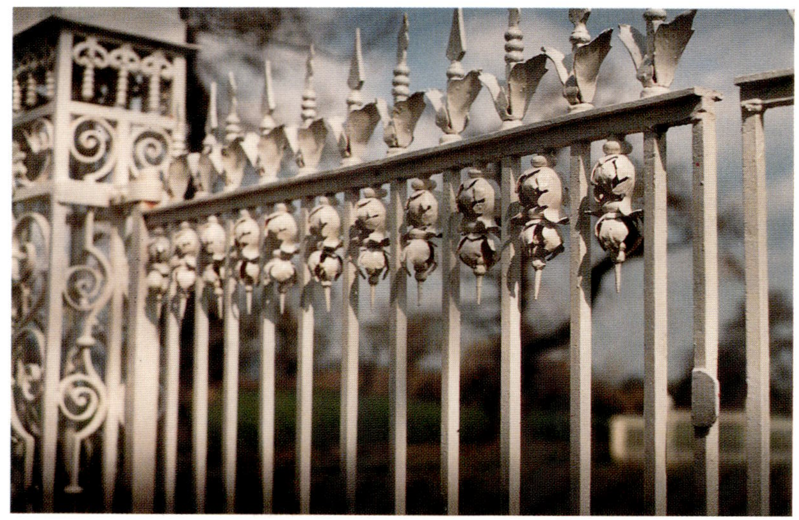

Cholmondeley Castle, nr. Malpas, Cheshire.

BAKEWELL AT CHOLMONDELEY

Perhaps the most surprising feature of Bakewell's efforts at Cholmondeley Castle is the single leg design of his balusters seen in his out-of-doors work such as the Chapel steps, the Moat garden and the Loggia. (Plate 8). All other Bakewell balusters (Fig. 3) have two points at which they are fastened into the steps. One can assume that they were used at Cholmondeley as a result of Tijou's influence and that Bakewell was not particularly enamoured of the idea; whether his objection would be constructional or aesthetic is not clear.

The Chapel Screen (Top left, Plate 8) is a delightful and delicate piece of work. The Pilasters are beautifully scrolled and are topped by neatly moulded caps. The dog bars (Fig. 4) are definitely Tijou derivatives and the pendant fringe (Fig. 7) of double husks is taken from Tijou's *A new Book of Drawings*, published in 1693. The upright fringe has alternating motifs; apart from this example , the only other occasion is in his fringe at Worcester. (Plate 15).

The Great White Screen, though deprived of gates, makes a most impressive entrance to the Castle grounds and is an excellent example of all the design ploys used in gatesmithery. (Fig. 1). "Reaching for the Sky" comes to life on Plate 8. (Also refer to Figs. 8 and 9). The upsweeping bead points and the climb to a high central point all maintain this feature. As the Marquess and Lady Cholmondeley open their gardens to the public on Sundays from April to October, there is no difficulty in viewing the screen.

COMBERMERE ABBEY, CHESHIRE AND OLD WESTBURY GARDENS, NEW YORK, AMERICA.

During the period of eleven years when Bakewell was engaged on work for the Cholmondeley Estate (1711 - 1722), it is highly probable that he made the gates for Combermere Abbey — a mere six miles distant. We know beyond any shadow of doubt that he made the gates because his RB signature was discovered upon them. A letter in *Country Life* Nov 29th 1919, written by John Starkie Gardner reveals that he had the task of restoring the Combermere gates; that from the start he thought the gates to be by Bakewell and furthermore that his wife had actually discovered the RB signature. Eventually the gates were re-erected at Westbury House, Long Island, New York, U.S.A.

The author is indebted to Miss Laura J. Carpenter, the archivist at the Westbury House, Old Westbury Gardens for the following potted history:

> Although the construction of Westbury House, begun in 1903, was essentially complete by 1906 when John and Margarita Grace Phipps and three of their four children moved in, the Bakewell gates did not grace the main entrance until 1920. In 1918 George A. Crawley, architect of Westbury House, came across the gates. Although the gates had fallen into disrepair, Crawley was certain that he had finally found the perfect pair for Westbury House and wrote to Mr. Phipps of his find. "I am having this remarkably fine pair of Old English Gates restored to a pristine and perfect condition — a work needing experience and skill .... as to the side railing they are practically scrap iron but will be used and rebuilt." J. Starkie Gardner of London did the restoration work for nine hundred dollars plus one hundred and fifty for painting, the Bakewell gates were shipped to Westbury, and in 1920 were installed as the entrance gates to John S. Phipps country estate - Westbury House.

In the frontispiece one can distinguish the cypher of J. S. Phipps. (See the Bibliography — *English Iron in a New York Garden*)

Cholmondeley Castle, Malpas, Cheshire.

The Great White Screen and —

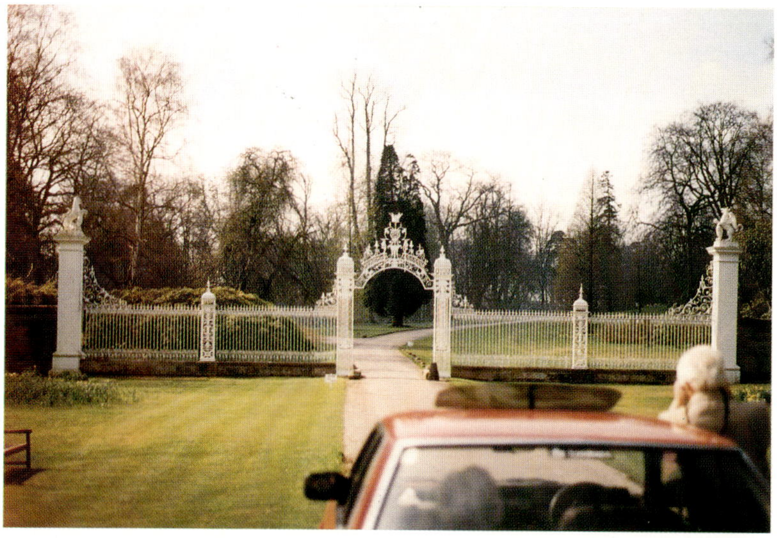

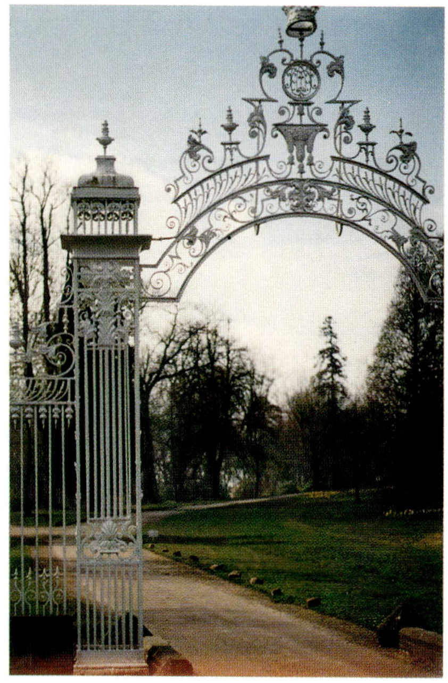

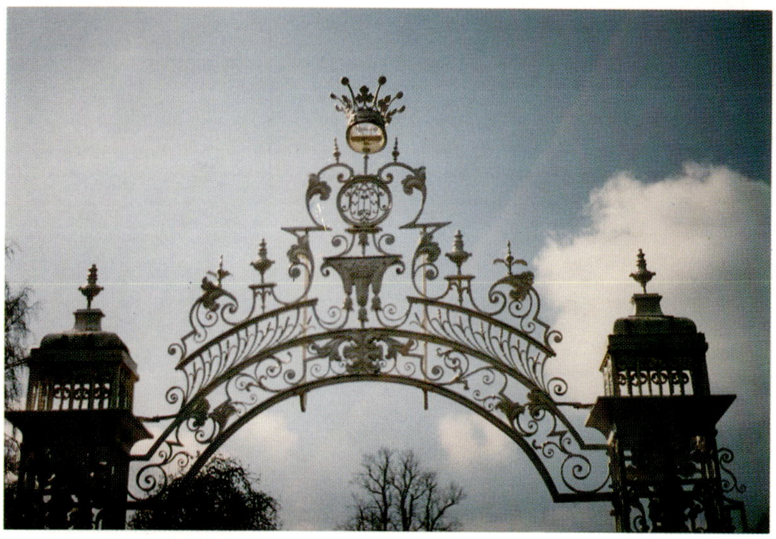

The Moat Garden

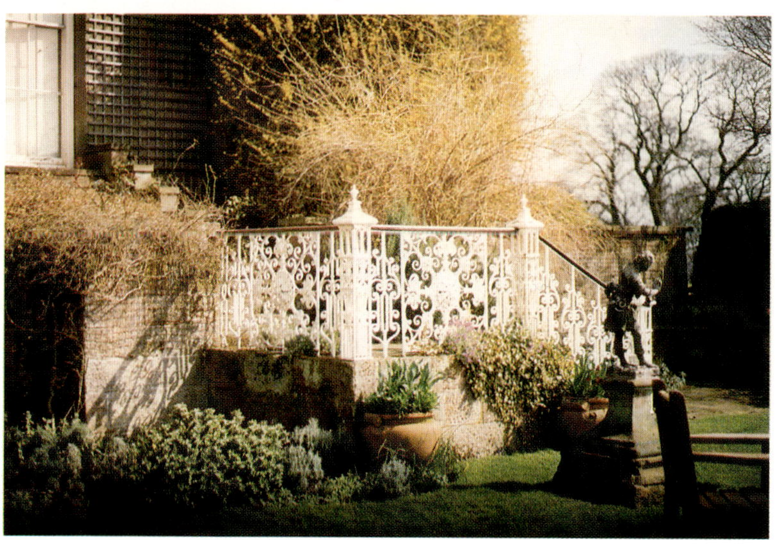

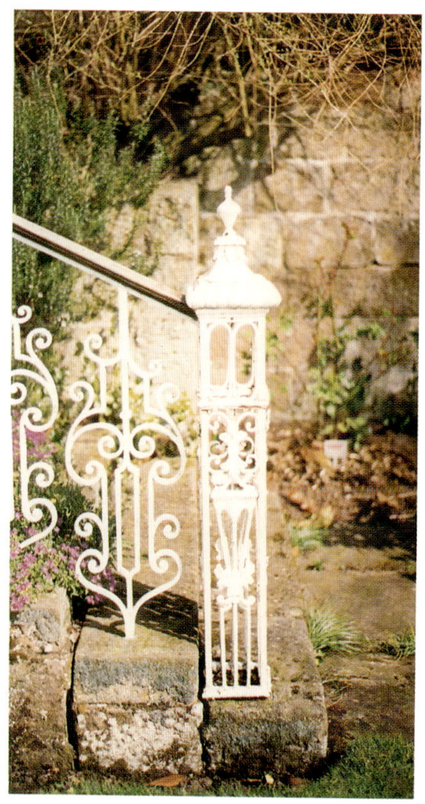

BAKEWELL AND THE EARL OF STRAFFORD

Deposited at the British Library is an Additional Manuscript, No. 22241 which is the original of an agreement between Robert Bakewell and the Earl of Strafford concerning the making of wrought ironwork at Stainborough Hall, near Barnsley. This Hall was later renamed Wentworth Castle. Bakewell referred to it as "Stoneborrow Hall".

The agreement reads as follows:

A memmorrandom of an agreement made betwixt The Rt Honrb ye Earle of Strafford and Rob Bakewell for ye makeing A pare of Irron Gates and Pallosadings for ye frunt of his Ld Ships House at Stoneborrow. The Proposols as folloeth viz ye princopol bars of ye Gates to be made of Inch and Halfe Square bars and ye fluted bars to be five eight parts of an inch thick and Inch and Qrtor deep ye ornaments slender and proporcionable to ye rest of ye worke for ye Pallosades as folloeth ye Pillastors to be fixt ten feet Distanc from middle of one pillastor to the other ye pallosades bar yt go between ye pilastors are to be seven eight part of an inch ye bars to ye pilastors to be inch and Qr The distance of ye bars to be five inches from middle of one bar to ye other ye height to be six feet to ye top of ye spike note ye Rob Bakewell is to set it up at his own charge Likewise ye carriage of it is to be found by Rob Bakewell and his Ld Ship is to find Led for fixing the whole work the Sd Rob Bakewell to Have a grote a pound one with another and to be brought within ye compass of two hundred pounds the throw frontspiece to be brought and ye three gates to be brought and set up by the first of march next and the whole to be done according to ye design givern to ye Earle of Stafford and seigned and seled by him butt by Reason of bringing ye charge of it to two Hundred pound Rob Bakewell is aloud to Dominnish any of ye bars or worke so that ye said worke be strong enought to with Stand the wether and Don in ye best and workmanlike manner. It is Likewoys agreed betwixt ye Earle of Strafford and Rob Bakewell that he shall furnish Irron Rales for and according to ye moddol Givern in to Seprate ye Garding from ye Tarris on ye South end of ye house and ye tarris from ye other end of ye house with Gates to open on ye same walk as ye Rales to open ten futt away and the Pillastors to have the Armes and Gartor upon them Ye said Rob Bakewell to be Alowed a grote pound and to be made six futt height and he to be pade for ye worke as he setts it up. He is to have a grote a pound for ye orniments.

Rob Bakewell

Dorect to Robt Bakewell Irron Gatemaker in Darby.

This Ironwork is no longer in being. Wentworth Castle later became a Teacher Training College.

BAKEWELL AND LORD ACTON

On October 6th 1718, there was an agreement concluded between Bakewell and Sir Whitmore Acton of Aldenham Park, Morville, Bridgnorth, Shropshire, concerning the making of two iron piers (page 44). This was a further agreement showing that work had been done previously and that these piers were to replace others, probably of stone. The two photocopies of the original memorandum and the accompanying receipts are self explanatory, apart from the word "darts". This may well refer to dog bars (Fig. 1), or to the flame tufts seen in the gates and panels in Plate 9. Also on Plate 9 one can see that the stretcher frame (Fig. 1) is perfectly flat as is the case on the West End gates of Derby Cathedral (Plate 24). The overthrow still manages the impression of reaching for the skies, helped as it is by the line of pseudo ellipses (Fig. 14) and the sweep towards the centre of the bead points. These underpin a climbing overthrow complete with both medallion and scrolly acanthus. Tables and flame tufts form the stepping stones to the high point. The spurred boot, the cloths of estate, the beads and water leaves all help to create a complete and dignified overthrow crowning a substantial centre piece. The pilasters show the strapwork effect of the divided diamond, the moulded caps the inset rectangle and other linear patterns. (Fig. 12). The dog bars are the customary arrow heads and as fringes there is a generous spread of one hundred and twenty flame tufts between the two stone pillars. The most unusual feature is the gap between the stretcher frame and the top rail of the gates.

PLATE NINE

Aldenham Park, Morville, Bridgnorth, Shropshire

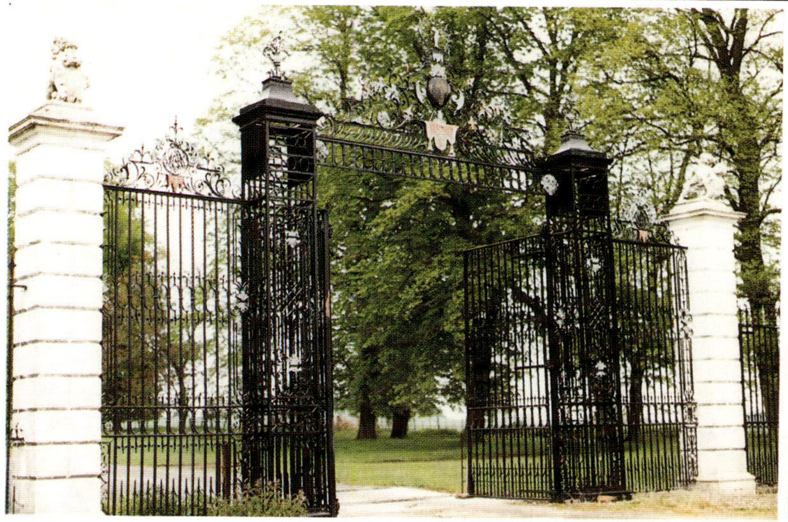

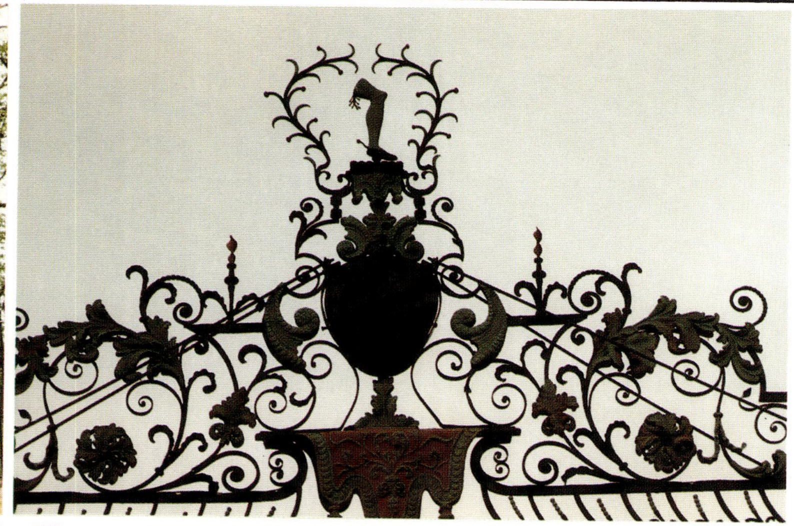

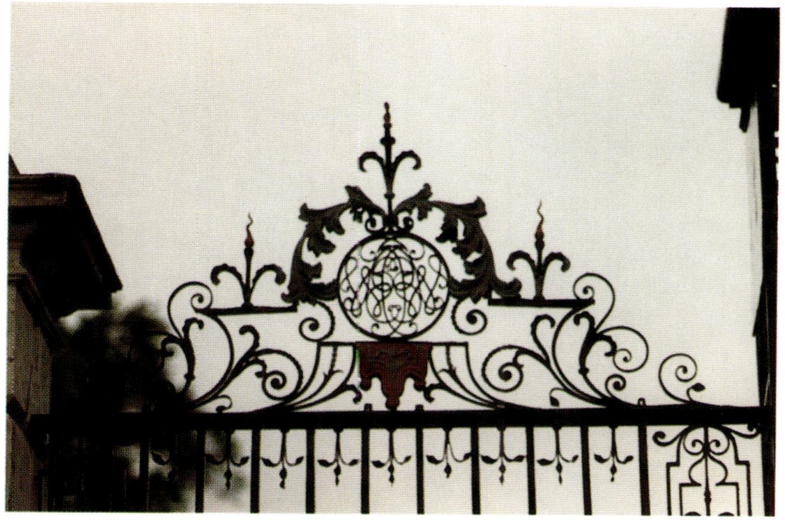

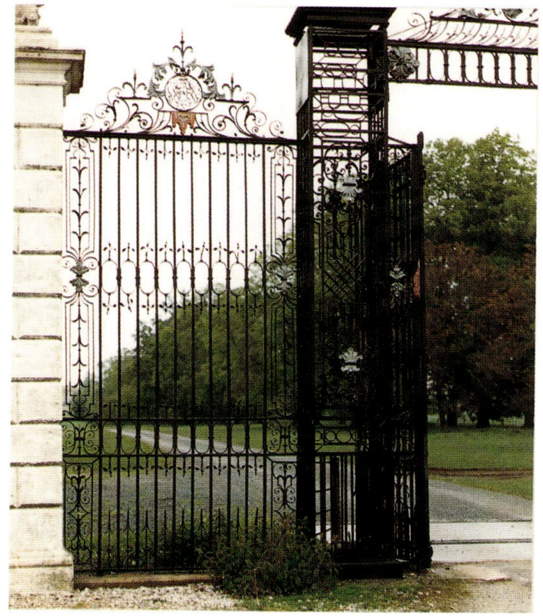

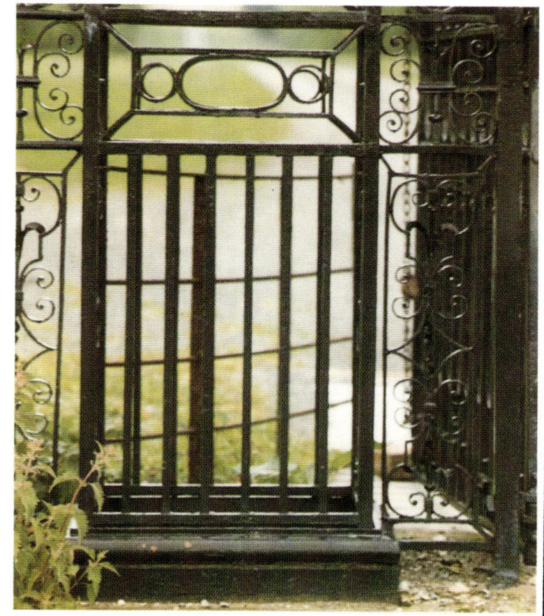

The Acton Gates. 1718.

BAKEWELL'S FONT COVER

Mr G. W. Beard, writing in *Country Life* May 2nd 1952, refers to a receipt discovered by Mr. Edgar Osborne and Mr. Francis Fisher at Radburne Hall when they were engaged on work for the National Archives. The receipt applies to payment for a font cover for the Parish Church of Saint Werburgh in Derby and reads thus:

> July ye 5 1718
>
> Received of Mr. Charles Benskin ye sum of twelve pounds in ffull for a pece of Irron Worke and Stone Worke done for ye font in St Worbors Church
> I say received p me £12 0 0.
>
> Robert Bakewell.

Mr Benskin was a member of the yeoman fraternity in the Derby area and was something of a benefactor. Later we will see his gift to another Church within the Derby area.

The piece of work referred to in the receipt is the font cover seen in Plate 10. It is a unique artefact, totally different from any other example of wrought ironwork. Standing a little over three feet in height it represents a tribute to early eighteenth century craftsmanship.

The skill required to make the horizontal rings and the vertical profiles with such placing and balance is immeasurable. When one adds the decorative features such as the acanthus in its various forms, the water leaves and the turned finials it becomes a labour of love rather than a mere geometrical exercise.

The centre picture on the bottom of Plate 10 is of a lectern surmounted by a Pelican rather than the usual eagle. The Church acknowledges several symbols of the Sacrament of Baptism; the Pelican is one and the Salamander as seen at Youlgreave is another. It is not surprising then to learn that the pelican was originally inside the font cover.

Saint Werburgh's was extended in the year 1896 and the font cover was found in a poor condition in a cellar. It was restored and created such an impression that a new font was ordered for it together with the rise and fall mechanism in the form of a chain and counterbalance.

Saint Werburgh's was the venue for the marriage of Dr. Samuel Johnson and was later used as the Church of the Derby Diocesan Teacher Training College. Now it is one of the growing list of redundant Churches. Whatever its future, it is to be hoped that the font cover and the pelican can be reunited, the bird replaced in its true home and the ironwork re-painted in a good blue and the leaves and finials gilded. That would be a happy outcome!

BURNLEY BALUSTRADE

Plate 10, Bottom right, is taken from a transparency supplied by the Towneley Hall Art Gallery and is included purely on stylistic grounds. Fig. 3 shows a number of balusters and most of them are equipped with the spur mentioned in the Figure. The spur is recognised as a Bakewell trade mark as is the wavy line that appears in the balusters and also in the bottom panel. Plate 10 shows (Top left) the font cover in position. (Top centre) the ease with which the lady can lift it. (Top right) a further view. (Bottom left) greater detail and (Bottom centre) the Pelican Lectern.

PLATE TEN

The Parish Church of St Werburgh. Derby. Font Cover. of 1718 and—

Townley Hall, Burnley. Staircase Balustrade.

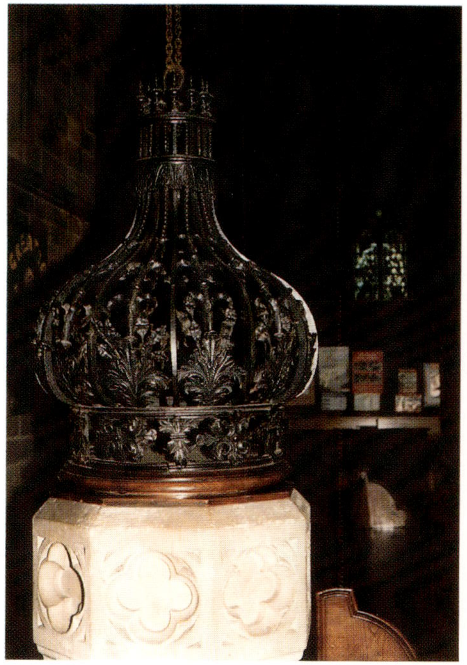

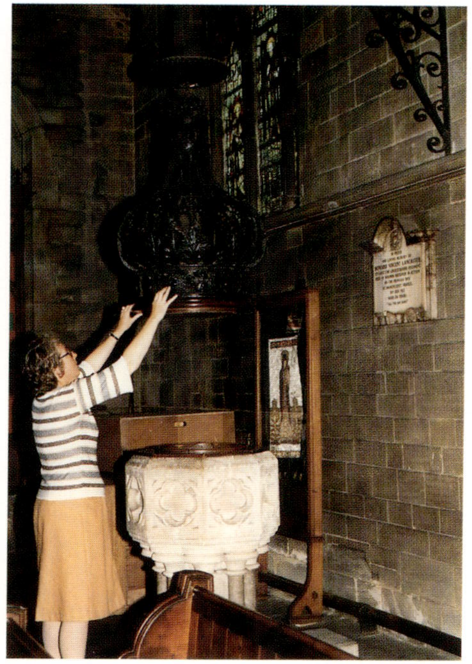

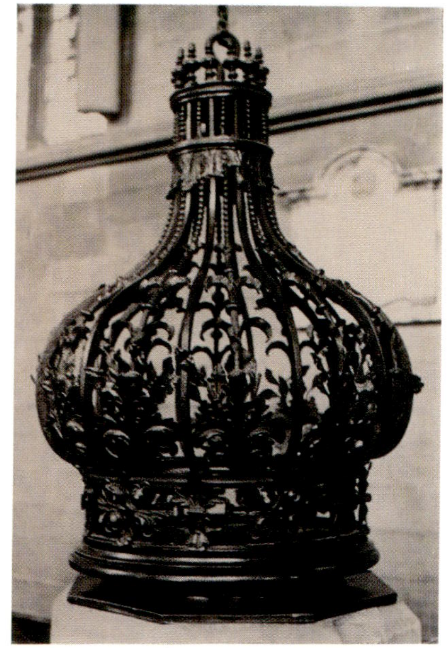

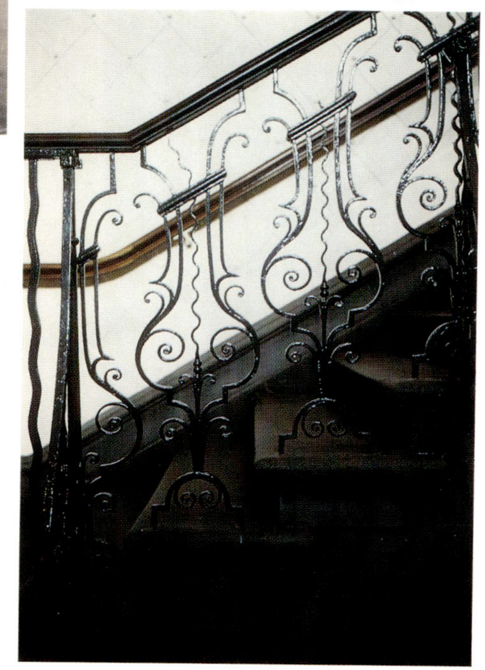

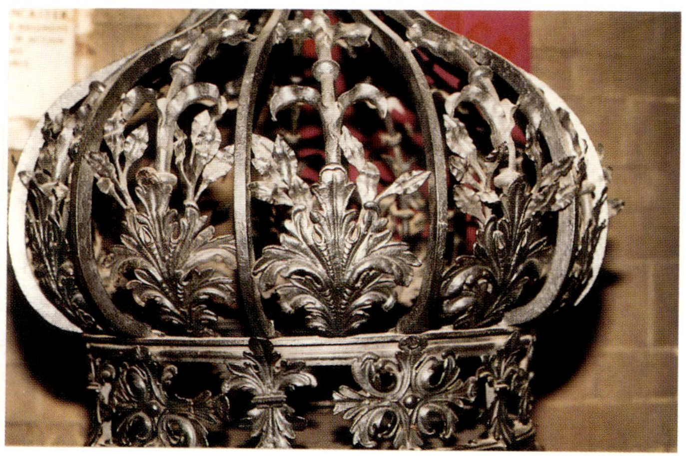

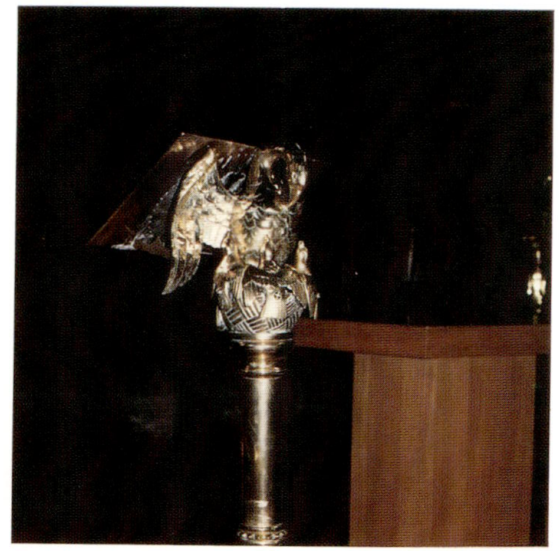

WEATHER VANES BY BAKEWELL

Calke Abbey, recently popularised as a time capsule when it was taken over by the National Trust, lies close to the village of Ticknall, near Derby. Howard Colvin, Fellow of Saint John's College, Oxford, who researched Calke and wrote the book *Calke Abbey, a Hidden House Revealed* also carried out a long process of preserving the muniments of the Harpur Crewe family. Amongst them are a number of Account Books, now lodged at the Derbyshire Record Office in Matlock (D2375/277/5 - 15). The first entry of interest in the context of Bakewell's work reads as follows:

> 1717. Pd Mr. Bakewell for a ffaine put up on the stables. £6. 0. 0.

Sir John Harpur was responsible for the creation of a new Stable Block, built 1712 - 16. Above the archway leading to the customary stable quadrangle is a cupola which was the position of the new Weather Vane.

In 1750, however, the Vane (or ffaine) required restoration so a further entry on the same item was required:

> 1750. Pd for the carriage of the vane and the ironwork for the Cupilo on the Stables to Derby and back. £0. 3. 0.
> Pd to Mr. Bakewell the Ironmonger for a new spindle and ornimental work to the Vane on the Cupilo at the Stables at Calke Ac as per Bill. £4. 5. 0.

In 1747 Sir Henry and Lady Caroline Harpur were responsible for building a Chinese House on a small island in one of the ponds at Calke. It was a kind of summer house or gazebo - a little bit of a folly - and was decorated with Dragons and other symbols. Chinoisery was by this time becoming popular in England. The Vane is well documented, but no longer extant. The following entry corroborates its former existence and was probably written by Sir Henry.

> July 1747.
> My expenses at Derby with Mr. Bakewell about the Vane for the Chinese House. 5. 0.
> December.
> 1747 Pd Mr. Bakewell the Ironmonger. £8. 0. 0.

SAINT ANDREW'S CHURCH, RADBOURNE

The Church is built close to Radburne Hall, the seat over many, many years of members of the Pole family and is host to a spidery but effective weather vane made in Bakewell's workshop and listed by Pevsner as follows: "Weather Vane by Bakewell, 2s 6d. undated, probably an apprentice piece". It is very plain, very simple but obviously much prized as it has graced the roof of the Church Tower for 250 years or more.

SAINT ALKMUND'S, DUFFIELD, NEAR DERBY

The Church spire at Duffield is very tall indeed and is crowned by a Weather Cock. Pevsner lists this artefact and dates it as 1719.

Plate 11.	(Top left)	The Cupola and the Vane at Calke Abbey.
	(Top right)	as above but greater detail.
	(Bottom left)	Saint Andrew's, Radbourne.
	(Bottom right)	The Duffield Spire and (inset) the Weather Cock.

Also in 1719 Bakewell made ornamental ironwork for a walled garden at Calke, now completely altered and the ironwork demolished. In the account books we read:

> 1719/20 Mr. Bakewell, the Remainder parts of his Bargaine for the Iron Worke in the Garden being in full of all Accts. £10. 0. 0.

PLATE ELEVEN

Weather Vanes.

Calke Abbey 1717&1750.
St. Alkmund's, Duffield 1719.
St. Andrews, Radbourne.
(undated).

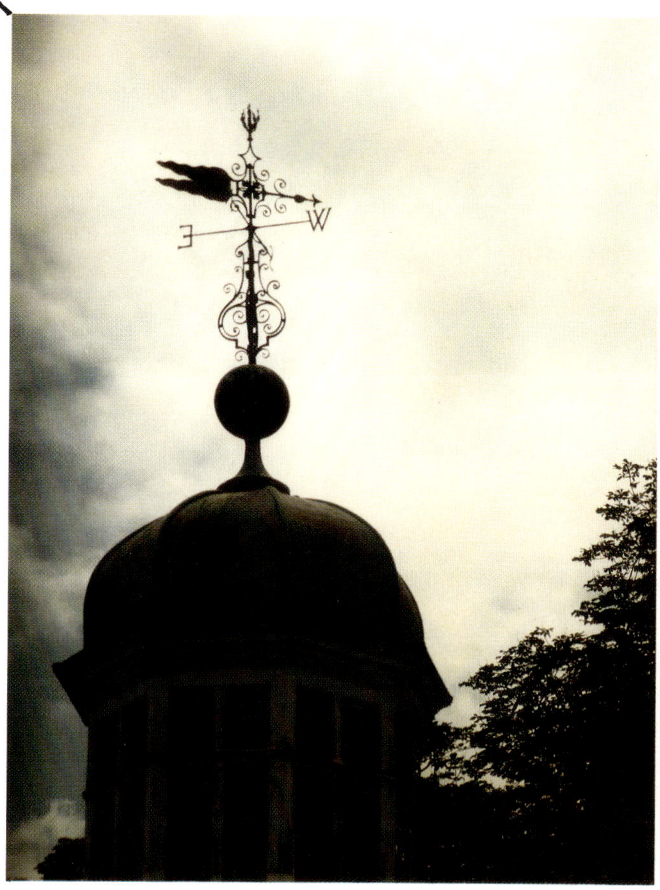

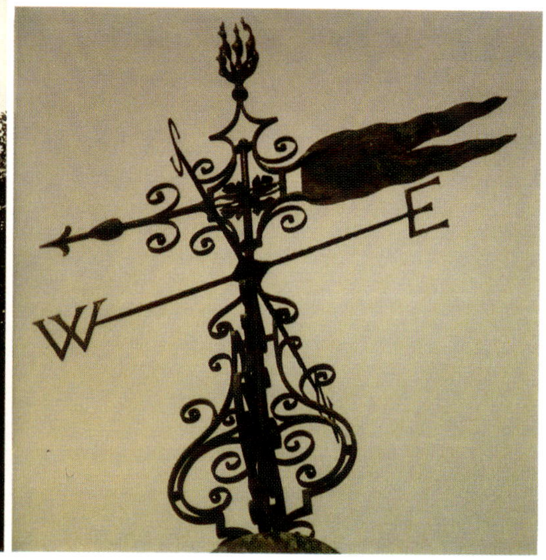

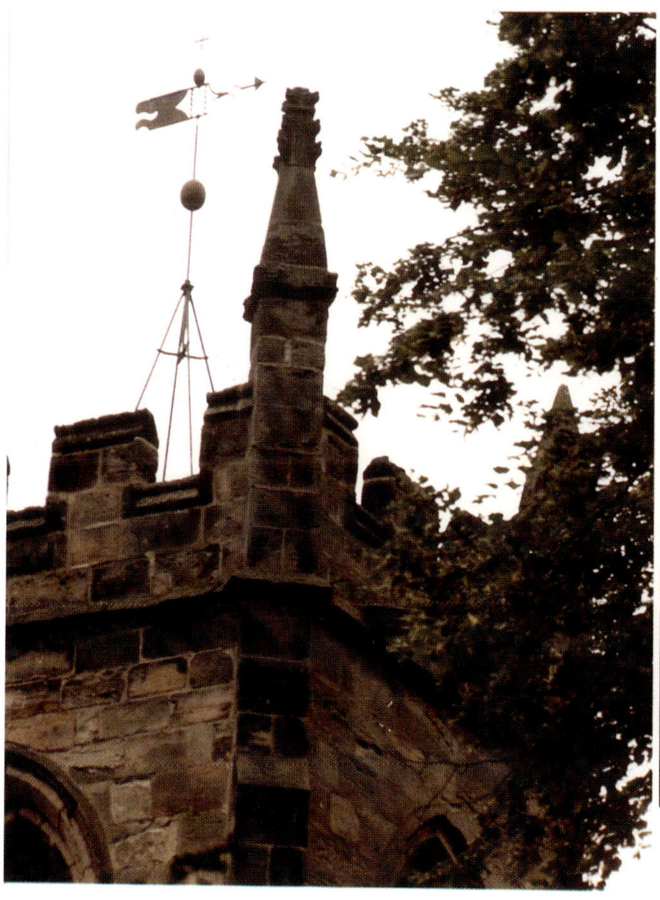

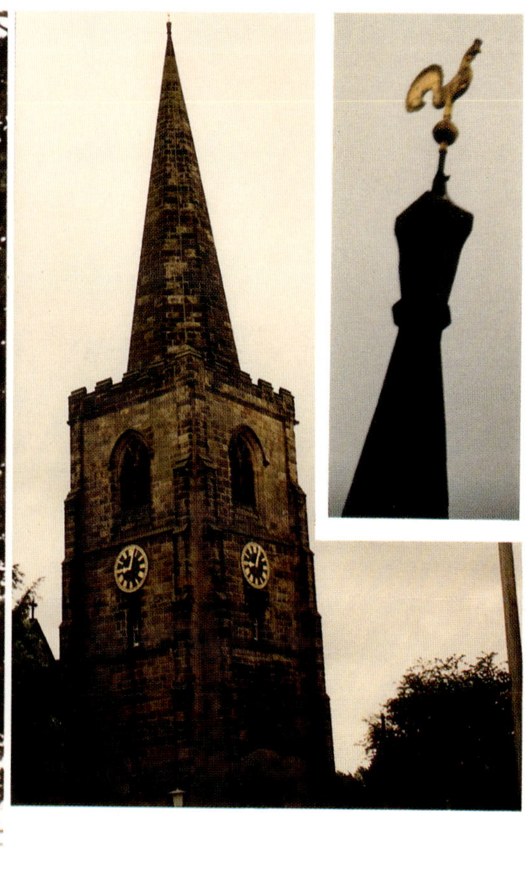

RAVENSTONE. Ashby de la Zouche and Coalville stand on the A50 in Leicestershire. Halfway between the two is a minor road leading South by South East to the pretty village of Ravenstone. This is host to a cluster of charming Almshouses and an equally charming Church, dedicated to Saint Michael. The outstanding piece of Church furniture is the Communion Rail which was installed in 1714. (Plate 12, lower picture). Bakewell's characteristic wavy line is present in a number of narrow panels composed of thunder and lightning rails. (Fig 4). There is a profusion of rosette or medallion acanthus and scrollwork decorated by the more mobile variety of the same leaf, water leaves and beads.

The gilders have been too eager, even treating the four masks as their territory, sedate masks which show that Bakewell could, after all, produce good work even in this line. The blue carpet on the steps complements the gold of the ironwork - what a pity that the kneeler cushions were not of the same colour. A busy but attractive screen, beautifully made and well cared for. One must assume that the smithy in Oake's Yard was employing an increasing number of apprentices and journeymen as the Ravenstone screen was made when Bakewell was still carrying out work for the Cholmondeley estate.

BORROWASH, DERBY. The A52, travelling east from Derby, passes through Spondon and, continuing from there, roughly divides the village of Borrowash from that of Ockbrook. In the village of Borrowash is the small brick Church of Saint Stephen. In the village of Ockbrook there used to be Hopwell Hall built in the style of Francis Smith of Warwick. Francis Smith was very active in the Derby area in the 1720s and 30s and he and Bakewell worked together quite frequently.

The Hall, which is no longer in existence, once was owned by Thomas Pares of Leicester. He it was who presented the beautiful wrought iron screen to Saint Stephen's, Borrowash. In volume 2 of *The Derbyshire Country House*, Craven and Stanley, in reference to Hopwell Hall state the following:

"Some Bakewell ironwork, formerly at the house is now in the Church at Borrowash". One half of the screen is shown on plate 12 and, stylistically speaking, verifies the statement. The inset rectangle with the wavy line is seen both horizontally and vertically and the scrollwork and acanthus speak eloquently of Bakewell.

Borrowash, Derbyshire,
St. Stephen's Parish Church.

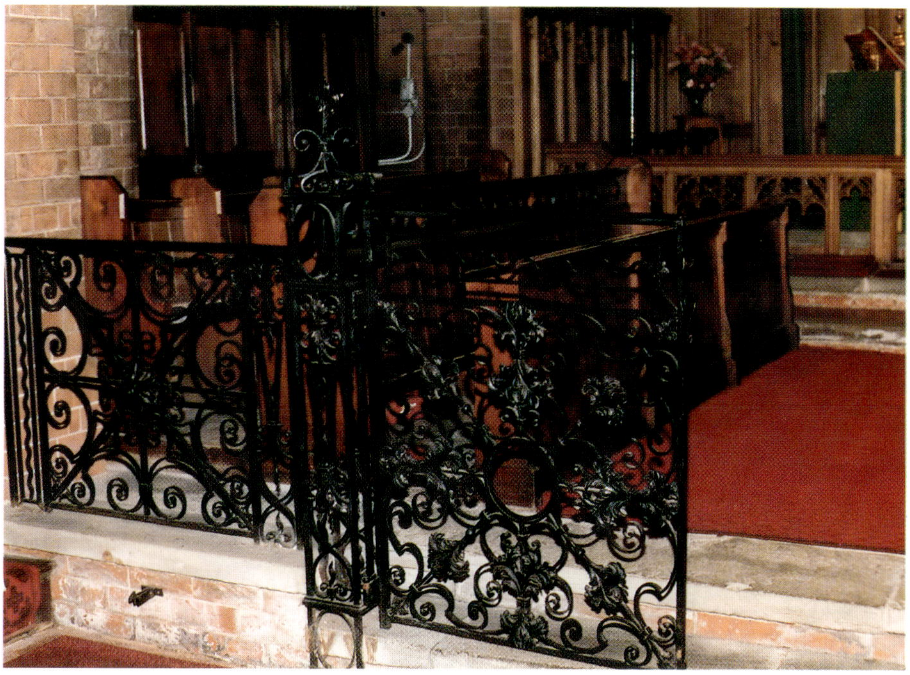

The Screen, circa 1723.

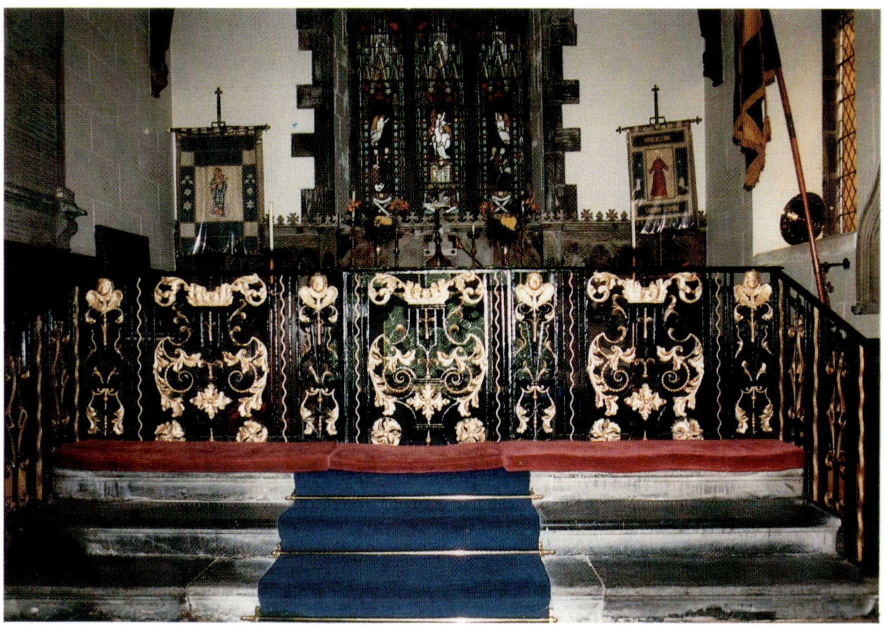

Ravenstone, Leicestershire,

St. Michael's, Communion Rails...

THE INDUSTRIAL MUSEUM, THE SILK MILL, FULL STREET, DERBY———

No more fitting place could have been selected anywhere as a venue for a Museum of Industry. It is housed in what is commonly referred to as 'The Old Silk Mill' which is the site of the very first factory in this country. Although its history is rather a chequered affair it is still on its original site and traces of its very first foundations can still be seen.

The mill was built in its present form following a fire in 1910, and became the Industrial Museum in 1974 since when it has housed many features of Industrial Derby, amongst which is a section on Robert Bakewell, the Derby gatesmith.

THE FANLIGHT. Top left and Centre left are two views of a fanlight grille known variously as the Darwin Fanlight, the Beaconsfield Fanlight and with much greater justification as the Heath fanlight. The upper picture shows the full lunette and compares closely with the roundel at Barton Hall (Plate 25). The centre picture is that of the cypher of John Heath. Fig. 10 shows quite clearly how a cypher is designed and uses this particular one as an example.

THE CHESSHYRE GATE. The Derby Oligarch, Gilbert Chesshyre, married into a branch of the Meynell Family. Sometime later, in 1705 or 1706 he received a bequest and this new found wealth enabled him to build No. 26 Friargate, a house of some pretensions in a road of handsome buildings. Custom has it that the gate by Bakewell (Plate 13 Top right) was a feature of this new house. The gate appears to be an exercise built around the inset rectangle (Fig. 6) as there are some thirteen examples of this motif, some of which contain the wavy line, some used as a vehicle for the medallion acanthus, and others completely free of any other motif. The Bakewell border runs from the lock upwards, around the lunette and returns to the lock frame on the opposite side. The ovals or pseudo ellipses, extended on their major axis by bead points are used to emphasise the vertical tendency. The fringe below the upper horizontal frame is reminiscent of Bakewell's work at Preston-on-Weald (Fig. 7). The lunette with its tables, scrolls, water leaves and beads surrounding the central cartouche are typically Bakewell but the crest, apparently, is of a Robinson, not the Chesshyre Family.

THE RIVETT ROUNDEL. This rather tired-looking remainder of a roundel is one of the examples shown in Fig. 10, which outlines the making of a cypher. It came from a house built in the seventeen thirties, for Alderman Rivett, a member of a yeoman family with connections at Repton and King's Newton. This house along with several others was demolished to make way for Corporation Street in Derby. It contained a certain amount of Bakewell work including this roundel which is seen at the Bottom right of Plate 13.

EATON HALL GATES AND RAILS. There is much controversy surrounding the making of these gates. There are two schools of thought; one is convinced that they were made by the Davies Brothers of Bersham, Wrexham; and the other believes that Robert Bakewell was concerned. Ayrton and Silcock in *Decorative Wrought Ironwork of the late Seventeenth and Eighteenth Centuries (1929)* state categorically that the Davies Brothers and Bakewell did work together at some time. Looking at the gates it is possible to see motifs regularly used by Bakewell. For instance the inset rectangle is at the base of the pilasters, at the top of them and in the side panels. Also in the pilasters is the strapwork diamond that Bakewell introduced on the pilasters at Aldenham Park. The panel roundels have much in common with that at Barton Hall. (Plate 25).

Many people take the view that no one man could have done all the work that is attributed to Bakewell. One must however realise that he had over forty years as a craftsman in his own right; that later he had a staff of apprentices and journeymen; and when he was overloaded with orders he could fall back on other craftsmen in other firms to carry out work to his designs. This last option could well apply to Eaton Hall Gates. They are shown on Plate 13 Bottom left and in the Bakewell section of the Museum.

The Industry Museum, Derby, The Robert Bakewell Section....

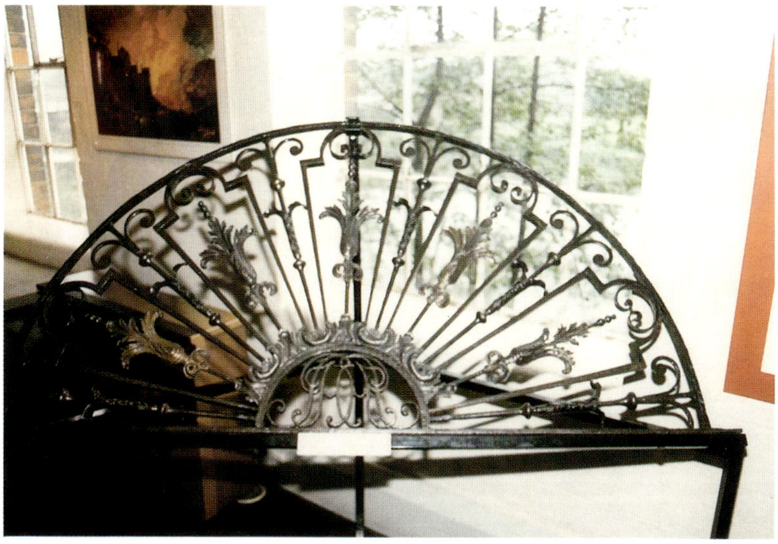

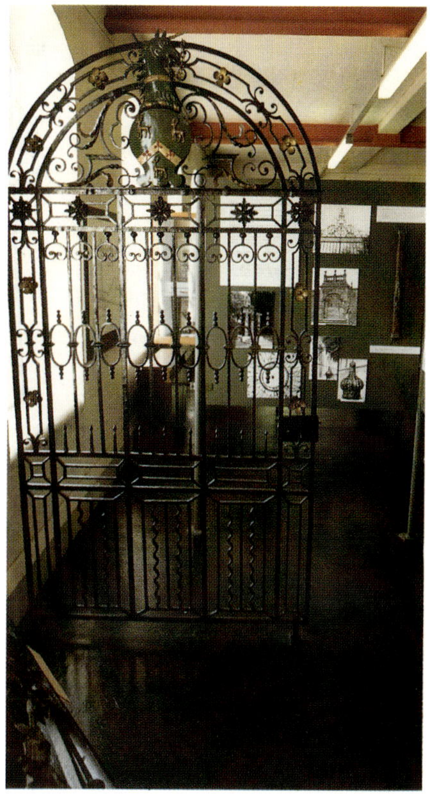

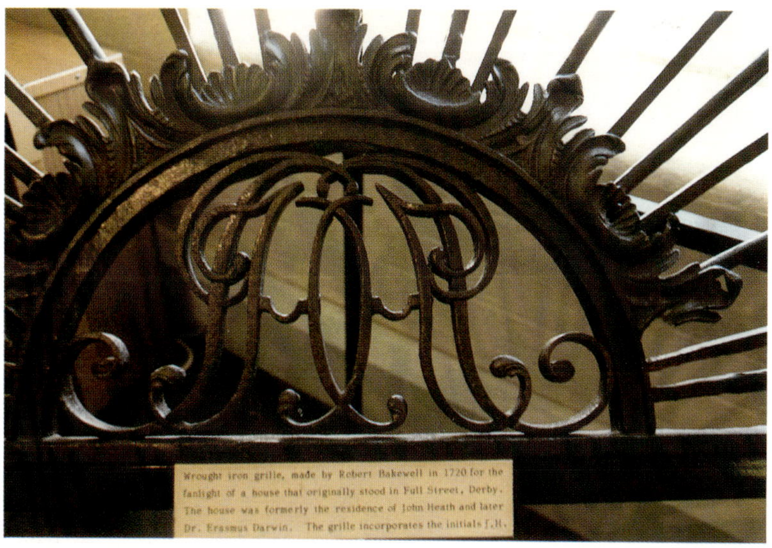

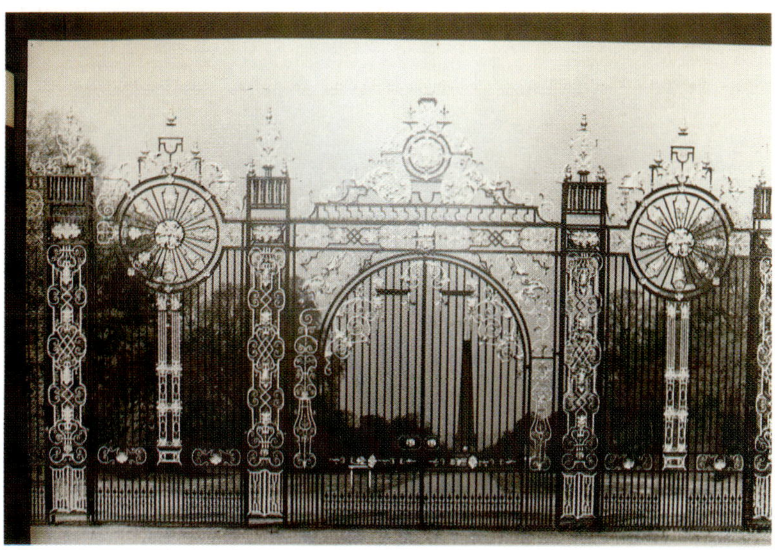

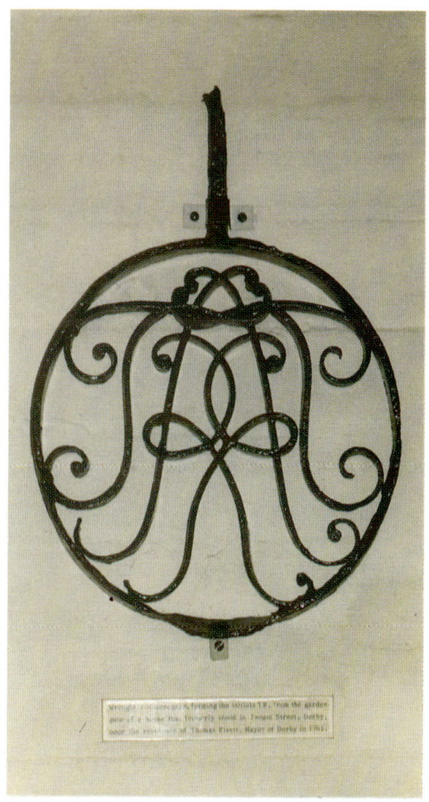

SAINT MICHAEL'S PARISH CHURCH, ALVASTON, DERBY

Alvaston is a suburb on the South East of Derby. Off the main road, the A6, is the Church of Saint Michael. On the East wall of the North Aisle is a wrought iron Reredos. Nikolas Pevsner in his *Buildings of England - Derbyshire*, refers to it thus: "an extremely pretty, if somewhat rustic, piece of wrought iron with a figure of an angel and lambrequins and scrollwork speak of the existence of the previous Church. (The ironwork is a reredos given by Charles Benskin and may be by Bakewell)". Roy Christian in his book *Derbyshire* says "Alvaston, off the London Road, still has a recognisable village street and its Church of 1856 contains a Saxon coffin lid and some attractive wrought ironwork, probably early Bakewell, that have survived from an earlier foundation". In the author's opinion there can be little doubt concerning the making of this reredos. Charles Benskin (Page 34) had paid Robert Bakewell for the Saint Werburgh's font cover. It would be inconceivable that he would commission work from another smith after the success of that piece of work. Incidentally a lambrequin is, in wrought iron terms, the representation of a small piece of drapery and, like a cloth of estate (which is rather larger) was moulded from a softer metal than the iron. The picture on the left of Plate 14 shows four lambrequins along the base, and the drapery that bears the letters would be classed as a cloth of estate even though in this instance it is applied to a piece of Church furniture. The medallion or rosette acanthus seen on the outer limits of the reredos, the scrolls and the cursive scrolly acanthus speak of Robert Bakewell. The Angel, presumably, is Saint Michael, to whom the Church is dedicated.

THE HALL, TISSINGTON, ASHBOURNE

Tissington is one of the most beautiful villages of Derbyshire. It is probably best known for its tradition of Well Dressing. Every year, the wells, which have been colourfully dressed are ecumenically blessed, several faiths being represented. There are six of these wells, one of which is directly opposite the Jacobean Hall, built in 1609. This well is dressed by Lady FitzHerbert with a band of volunteers from the Hall. Plate 14, Top centre, shows a gate surrounded by heavy stonework. This cannot be classed as a wicket gate since it will not allow the passage of a man on horseback. It is classed as a garden gate. Though the purpose of the gate may appear to be mundane, the gate itself is not. The spear point dog bars, the inset rectangles seen in the lock frame and the side frames lead one to think seriously of Bakewell, whilst the row of ovals or pseudo ellipses remind one of the Chesshyre gate, the Aldenham Park overthrow, the Okeover overthrow (Plate 31) and the Melbourne arbour (Plate 2 and Fig. 14). Notice how close Bakewell came to making the narrow side panels into Corinthian Capitals, by his use of the acanthus. The Hall is visible in the background.

THE TABLE FRAME

This table has a gritstone top and a Bakewell underframe. The ironwork is finished in its original colours. It is documented as by Bakewell and its scrolly acanthus, its water leaves and its scrollwork are indicative of the craftsman. As this is in private hands there is little more the author can add.

Saint Michael's Church Alvaston, Derby. Reredos, & Table frame in private hands.

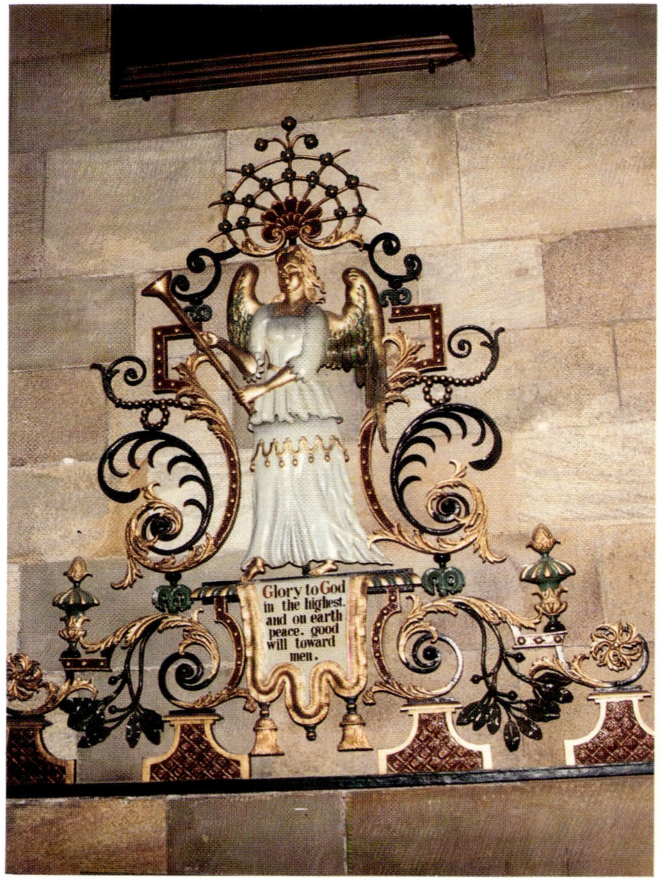

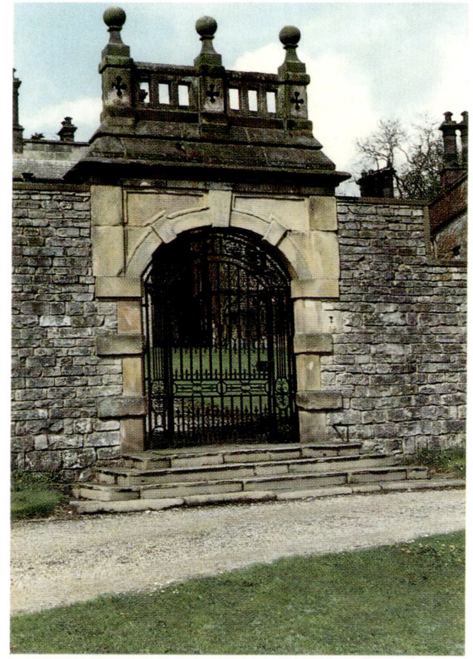

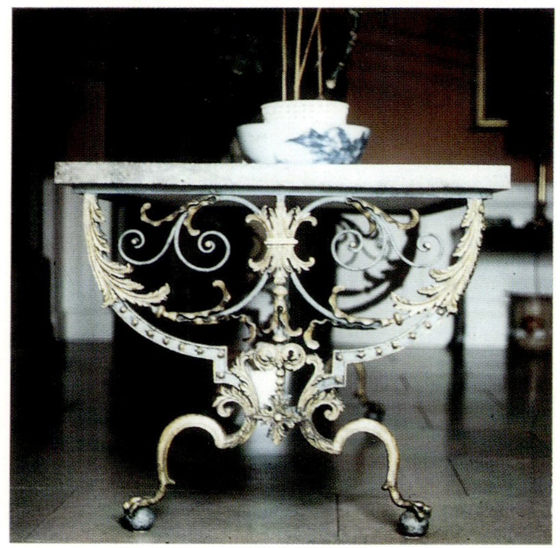

The Hall, Tissington, nr Ashbourne, Derbyshire, Bakewell's Garden Gate.

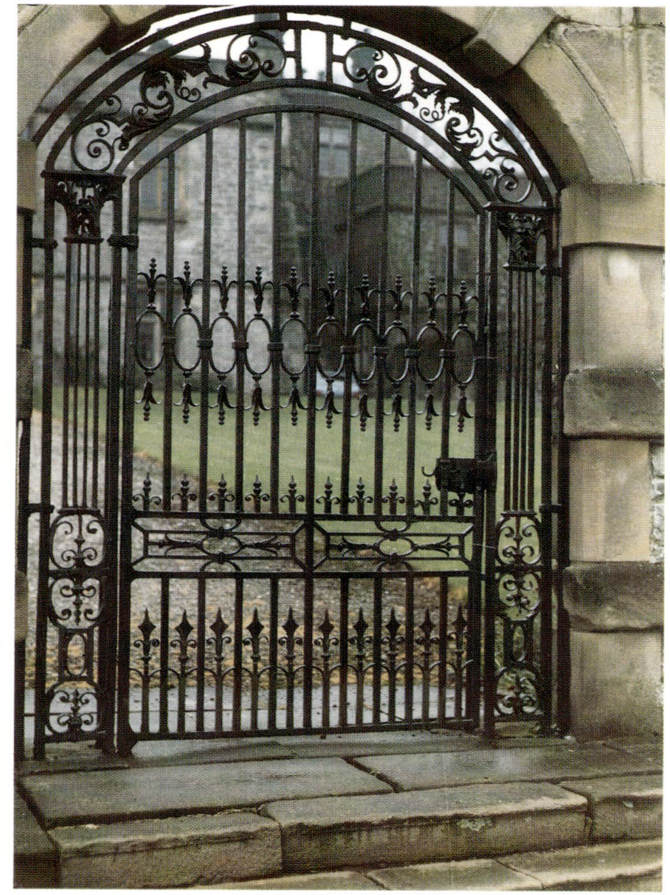

THE GUILDHALL, WORCESTER

To take a walk down the precinct in Worcester for the first time, particularly if the day is sunny, is an unforgettable experience. One is suddenly confronted by a remarkable blaze of colour as one approaches the Guildhall and its Bakewell ironwork.

John Starkie Gardner in *English Ironwork of the Seventeenth and Eighteenth Centuries* refers to these gates and rails. According to Gardner they were made by Bakewell in 1721 to 1723.

The Guildhall was the design of Thomas White who was a sculptor rather than an architect — he was actually awarded the freedom of the City of Worcester for his Statue of Queen Anne and in his will described himself as "Thomas White, Statuary". Examining Plate 15 one can appreciate that the work is Bakewell's as so many of the motifs he used are present. The inset rectangle used as a vehicle for the rosette acanthus, and again, for the ellipse; the Bakewell border in the stretcher frame, the 'finger board' construction (Fig. 11) seen in the pilasters and the use of the mobile and ubiquitous acanthus. Incidentally the pilasters are nine feet high and are topped by unusual caps of scrolled ironwork.

Reverting to Plate 15 we see:

Top left	The general view, most colourful
Top right	A pilaster, Note the scrolly cap.
Centre left	The overthrow. See Fig. 1.
Bottom left	A terminal bracket.
Bottom right	A standard. Note the finger board motif.

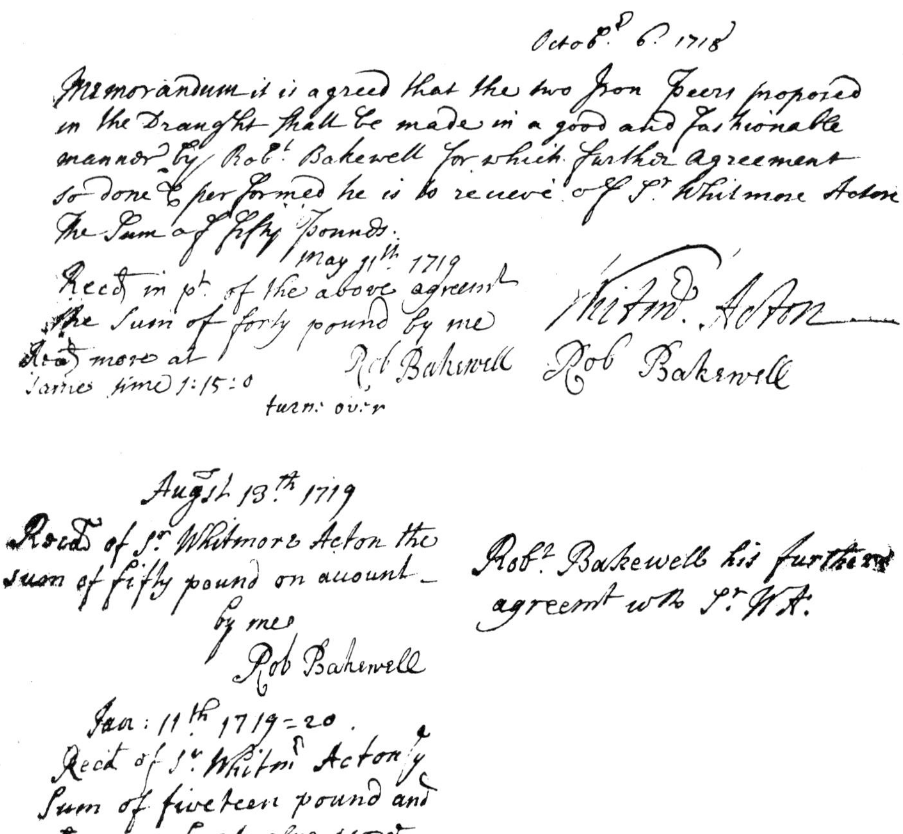
Octob.r 6. 1718
Memorandum it is agreed that the two Iron Peers proposed in the Draught shall be made in a good and fashionable manner by Rob.t Bakewell for which further agreement so done & performed he is to receive of S.r Whitmore Acton the Sum of fifty Pounds.
May 11th 1719
Rec'd in p.t of the above agreem.t the Sum of forty pound by me Rob Bakewell
Rec'd more at James time 1:15:0
turn over
Whitm. Acton
Rob Bakewell

Augst 13th 1719
Rec'd of S.r Whitmore Acton the sum of fifty pound on account by me Rob Bakewell

Rob.t Bakewell his further agreem.t wth S.r W.A.

Jan: 11th 1719-20.
Rec'd of S.r Whitm. Acton y.e Sum of fifteen pound and 12.s more for twelve score Darts by me Rob Bakewell

The agreement with Lord Acton (see page 32) and Bakewell's receipt.

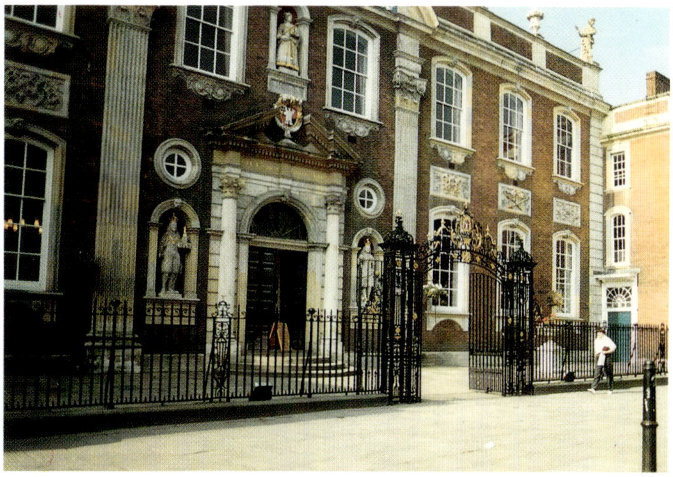

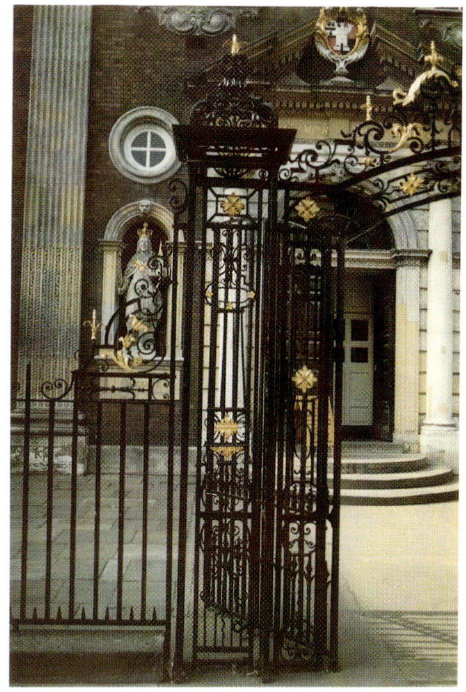

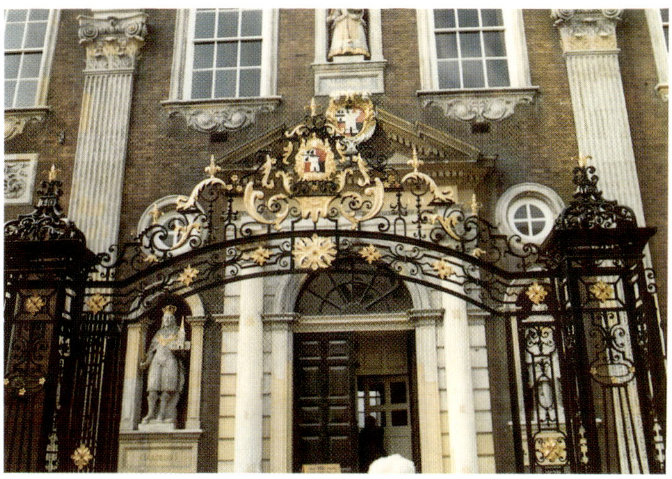

Worcester,
The Guildhall,
Gates and Rails.

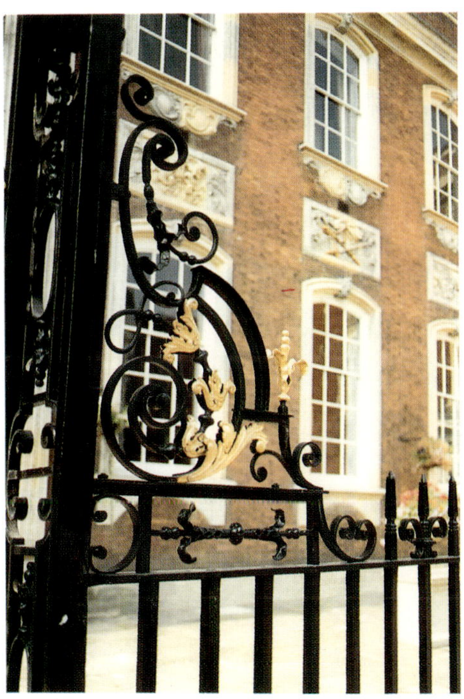

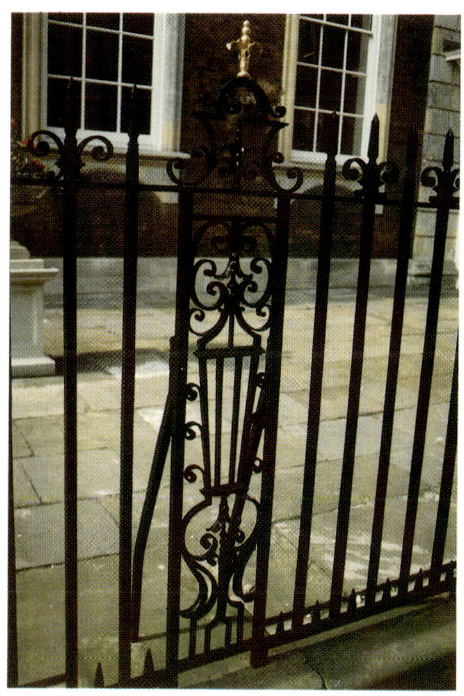

1721–23.

1722 was the year in which the 'Cavendish' Screen was started in what was then the Parish Church of All Saints. Dr. Hutchinson, the then Vicar, was responsible on his own initiative for the demolition of his Church. By subscription he managed to raise the funds required to restyle it. To prepare for the new building he hired James Gibbs, an architect with some Italian training, (who, incidentally was the architect for Saint Martins-in-the-Fields in London). The Smiths of Warwick were to build the new All Saints and Robert Bakewell was to craft the Chancel Screen. There is no claim that the present screen in its entirety is Bakewell's work. What is claimed for Bakewell is the centre piece between and including the main pilasters and the North gate to the Chancel. The remainder, beautifully in tune with the original was made by another Derby blacksmith, Mr Haslam, in 1904.

The Church is now the Cathedral and whilst it is architecturally sound and interesting, there is a bareness about it alleviated mainly by the astute use of colour and the splendour (not too strong a word) of the screen. It vies with the Melbourne arbour for the right to be termed Bakewell's Masterpiece! The screen was completed in 1725, as were several other artefacts which are documented by Bishop R. S. M. O'Ferrall in his *Pilgrim's Guide to the Cathedral, 1953.* Many Church screens are criticised because they obscure the East end of the building; the Cavendish screen does not, and the gates are rarely closed.

Payments to Bakewell for the screen and other artefacts are to be seen in the *Accounts of the Collegiate Church of All Saints' Derby* by Dr. Charles Cox and Mr. St. John Hope 1881.

Dr. Cox writes "several details of the interior will be spoken of in subsequent chapters, but the beautiful screen of wrot iron is too essential a portion of the fabric to be included in any account of the fittings... the screen was the work of a native of Derby, Mr. Bakewell for which he received £338.10. 0".

Nikolas Pevsner in his *Buildings of England* refers to the chancel screen thus: "The most important possession of the Cathedral, wrought ironwork by Robert Bakewell, the brilliant local smith who died in 1752". Raymond Lister says of it "Altogether it is a most impressive piece of work — it is said by some to be his masterpiece -but probably the most vital thing about it is its lightness which seems to force the design upwards to culminate in a mass of scrolly foam".

The book *The Cathedrals of England* by Batsford and Fry contains the sentence "There is call for much thankfulness that much of the Renaissance Ironwork of the great local smith, Robert Bakewell, remains in the Church".

Arthur Mee who wrote *The King's England* in 1937 in reference to All Saints' Cathedral writes "But the architecture is redeemed by a magnificent iron screen extending on either side of the Chancel to the side Chapels, its delicate scrolls and foliage as beautiful as anything Robert Bakewell made. It is a monument of fine craftsmanship." The best tribute comes from *British Architects and Craftsmen 1600 - 1830* by Sacheveral Sitwell: "Individual and beautiful ironwork which, anywhere but in our land of understatement would be as famous as the grilles and balconies of Jean Lamour in the Place Stanislaus at Nancy, is to be found at All Saints' Church in Derby now the Cathedral, from the hand of Robert Bakewell".

Plate 16	Top left	The main centre of the screen.
	Top centre	The inset rectangle and the wavy line.
	Top right	More inset rectangles.
	Bottom left	The North gate to the Sanctuary.
	Bottom right	Work by Mr. Haslam of Derby (1904).

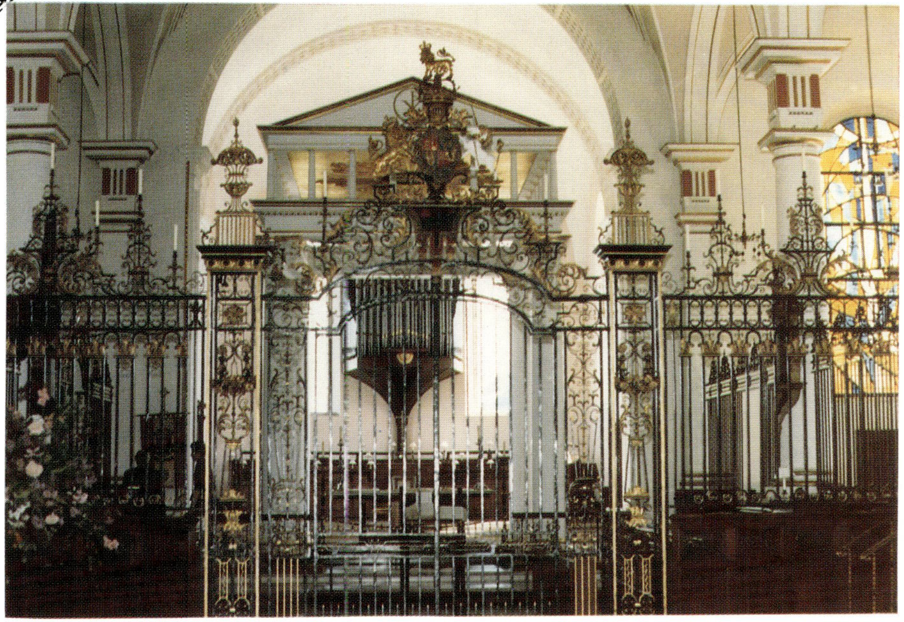

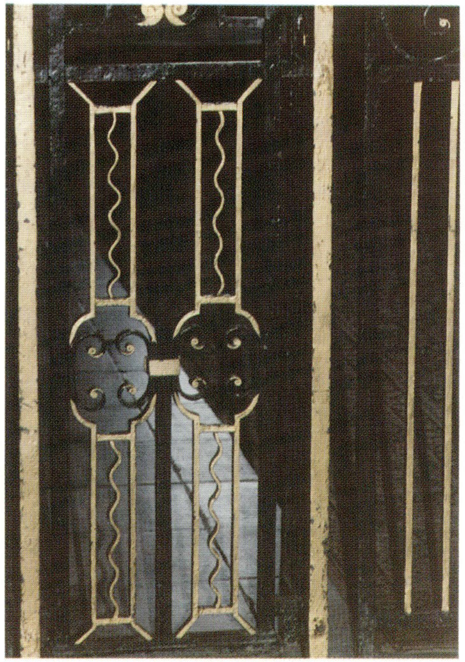

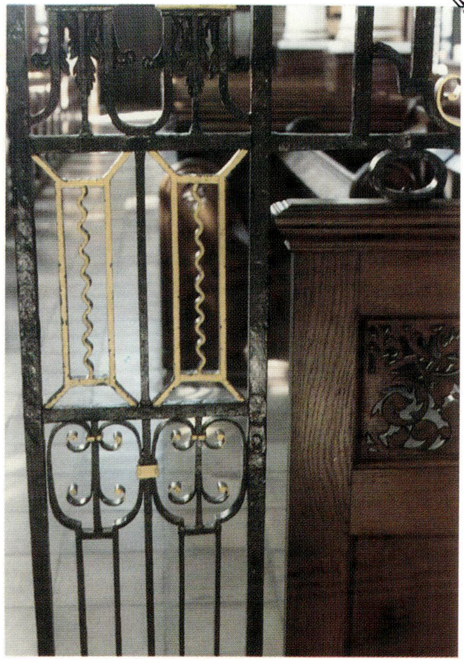

The Cathedral Church of All Saints', Derby. The 'Cavendish' Screen and some details..

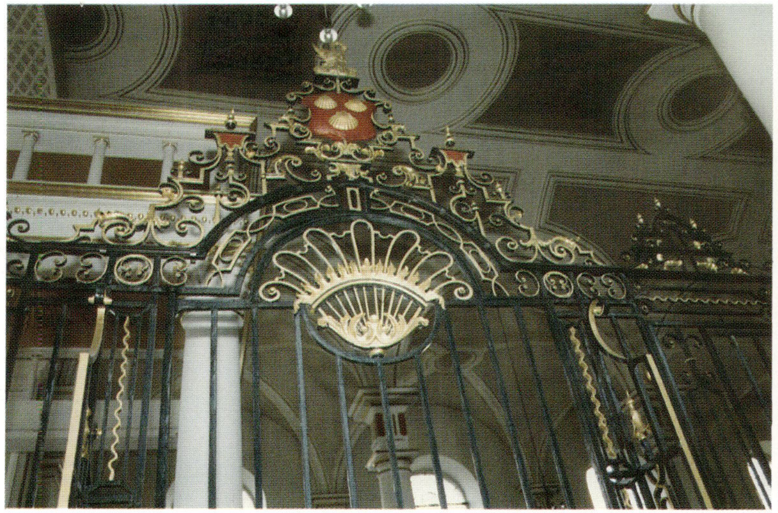

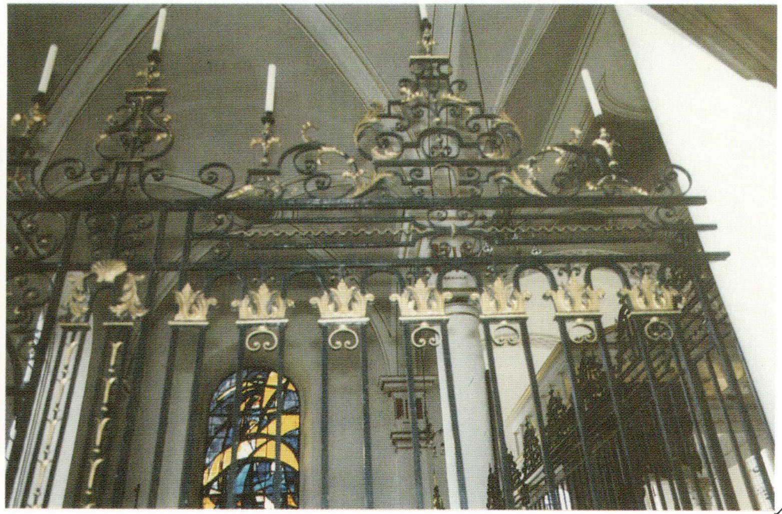

CHURCH FURNISHINGS, THE CATHEDRAL CHURCH OF ALL SAINTS

In his Guide Bishop R.S.M. O'Ferrall points out that the Communion Rails, the Consistory Lamp bracket, the base of the marble Altar and the ironwork of the Corporation Pew were made by Robert Bakewell in 1725. There is no acanthus anywhere superior to that on the Altar base. At the centre is a double motif, whilst at each end of the front and the forward part of each end the single motif gives a swirl and a movement which would be difficult to better. The corner brackets with their tables and water leaves, and the lines of husks in the spaces of the corner legs are exceptionally good examples of Baroque artistry in iron. The gilding, so generously applied emphasises the skill necessary to achieve such perfection.

The pageantry of City and Town affairs is occasionally taken into Churches and Cathedrals when Lord Mayors and Mayors, Councillors, Magistrates, Chief Clerks and other minor functionaries parade to services connected with Civic celebrations - one thinks of Commemorations, Hospital, Armistice and Law Session affairs. When the Sword of State is carried it is laid up during the service on a Sword rest. The Corporation Pew in the Cathedral caters for this and also for the Mace and the Lady Mace. Most of the ironwork is by Bakewell, but there have been one or two additions which, incidentally, harmonise particularly well. The use of the acanthus leaf and the cloth of estate beneath the medallion are particularly successful. The Buck in the Park is the true emblem of the Derby municipality. (The Derby Ram popularised by the local football club has no official recognition). In order to preserve the balance of Civic Pride there is also a County Pew which stands opposite the Corporation Pew. This is far more modern and designed as a tribute to the working men of the County.

THE COMMUNION RAILS. The rails at Foremark and Saint Ann's, Manchester are designed on the panel principle whilst the Derby rails are designed on the baluster idea. A number of balusters are placed in a frame together to form a unit. (Fig. 4). As is usual with any of Bakewell's balusters the most prominent feature is that of the lyre form. Scrolls and water leaves abound and the addition of gilded flame tufts (Fig. 13) adds further interest to the complete picture. The mahogany rail which lies along the baluster frames is smooth and warm and neatly moulded.

THE CANDLE SCONCE. North of the Cathedral Sanctuary is the oldest piece of woodwork in the building. This is the Consistory which was installed in 1634 and moved to its present position in 1725. It is, in fact a Seat of Judgment where complaints and suggestions are aired and judgments pronounced. It is on this woodwork that a candle sconce can be found. Rarely would one find so small an artefact endowed with so much acanthus. Bakewell, who installed it in 1725, must have revelled in decorating so simple a frame, of very few scrolls, with so much beauty.

In a building of Wren style, plain to a degree, relieved of itself only by the subtlety of colouring, it is good to see the work of the genius Bakewell adding strength, colour and diversion, not to mention charm to what is, after all, an architectural success.

PENSHURST PLACE, TONBRIDGE, KENT

In 1726 Francis Smith of Warwick built Wingerworth Hall, the hub of a most charming area, one-and-a-half miles on the Derby side of Chesterfield. The gates and rails were made by Robert Bakewell who frequently worked with members of the Smith Family. If anything could be said to be 'typically Bakewell' it would be this fine set of gates. Present is the Bakewell border, the spear point dog bars, the wavy line in plenty, the lock rail complete with bead points and the open bar construction of the vase form cap on the pilasters. Add to all this

Church Furnishings – All Saints' Cathedral, Derby. Altar Frame, Corporation Pew, Communion Rails and –

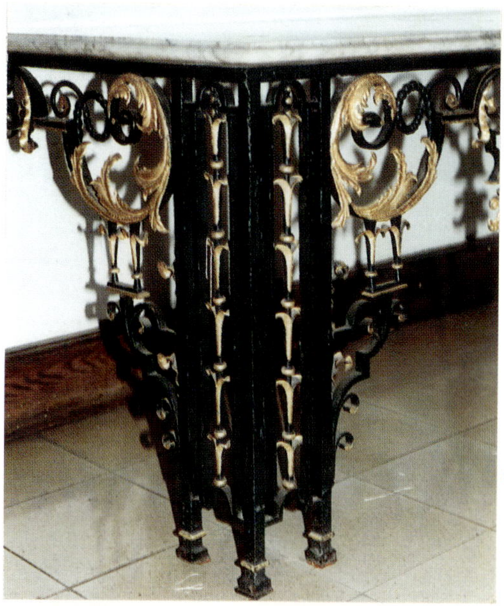

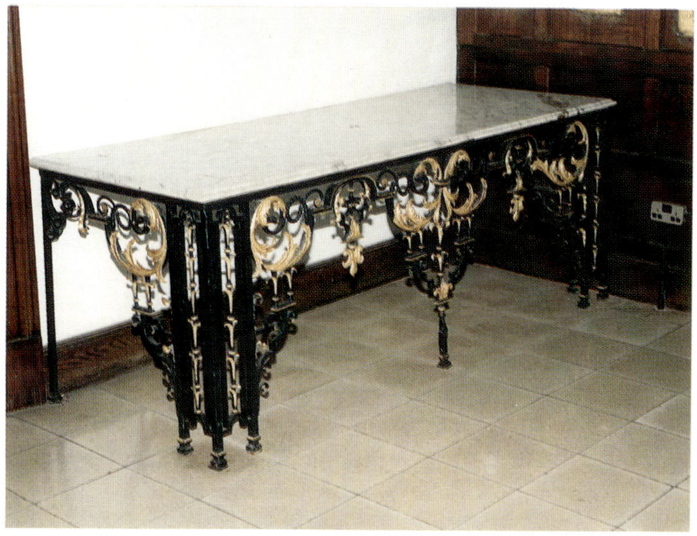

Candle Sconce.

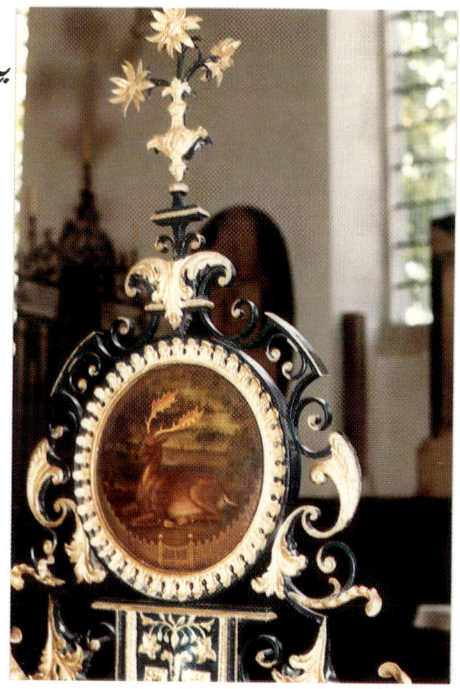

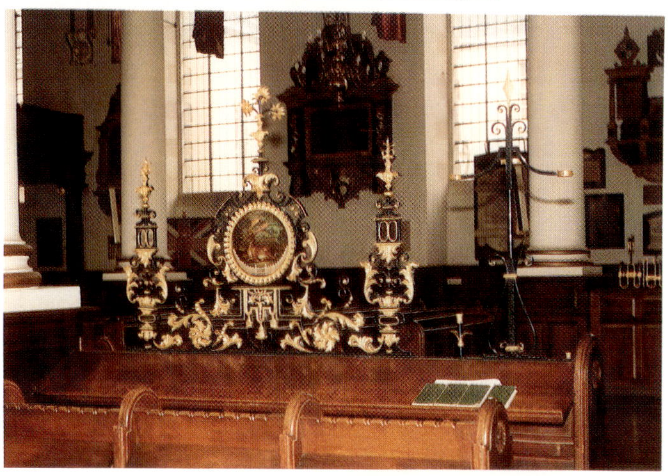

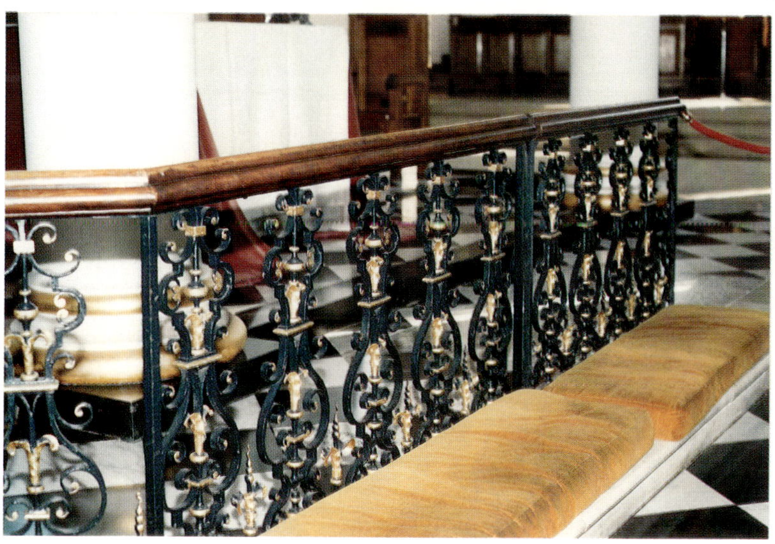

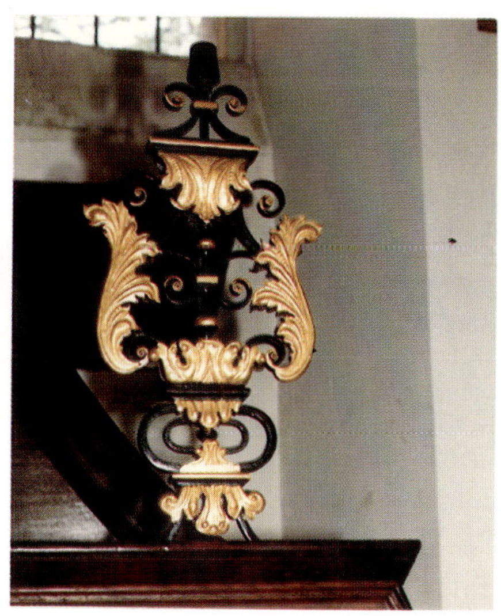

the tables forming stepping stones to the high points of the overthrows, complete with swirling acanthus and you have a potted demonstration of the skills of Bakewell. During a restoration some years ago the RB mark which is the undoubted proof of his work was found on one of the transoms.

Unfortunately alterations occur; the valley between the charming areas of Wingerworth and Hasland is now bedecked with a massive Coal Carbonisation Plant which the vandals of the Coal Board installed in the late fifties, destroying even the sweet pastoral smell of the area.

The Hall is demolished and the gates have had two important moves. First they were taken to a former Royal Residence, Savile House, Leicester Square, London, and then to Penshurst Place, having been purchased by the First Lord De Lisle and Dudley. (Grandfather of the present Lord).

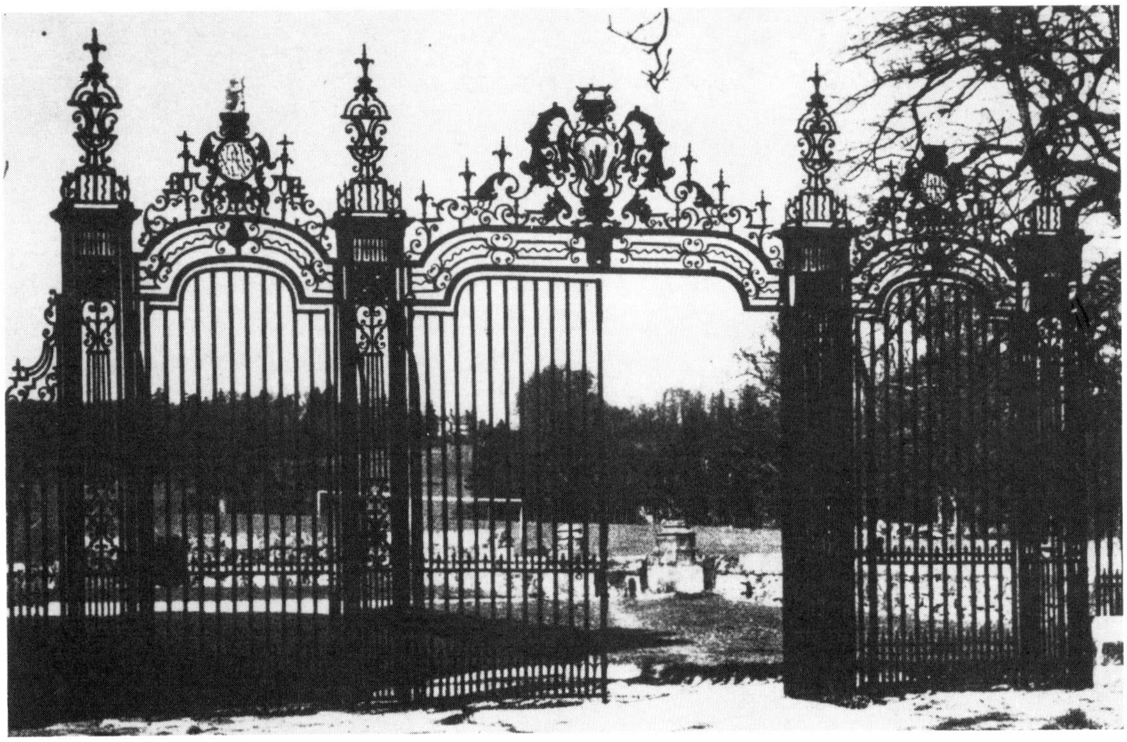

The entrance gates to Penshurst Place. These gates formerly graced the entrance to Wingerworth Hall, Chesterfield, and a royal residence in Savile Square, London.

DIGNITY AND DISTRESS, LLANHRYDD HALL, NEAR RUTHIN

Early in the 1980s the gates suffered the indignity of being badly damaged by a reversing dust cart. In 1987 an engineer carried out a partial restoration-cum-alteration. The overthrow, falling the furthest, suffered rather more than the rest of the gates with the result that the centre piece is missing; hence the now less than Baroque replacement.

Originally the Llanhrydd gates were very similar to the main centre of the Penshurst gates, being based on the raised stretcher frame (Fig. 9) and showing the bead point fringe, the lyre form and the Bakewell border, all adding weight to the claim that Bakewell was either the maker, the designer or both.

Although these gates have not wholly regained their former grandeur, they are, at least, possessed of a little dignity.

Penshurst Place and Llanrhydd.

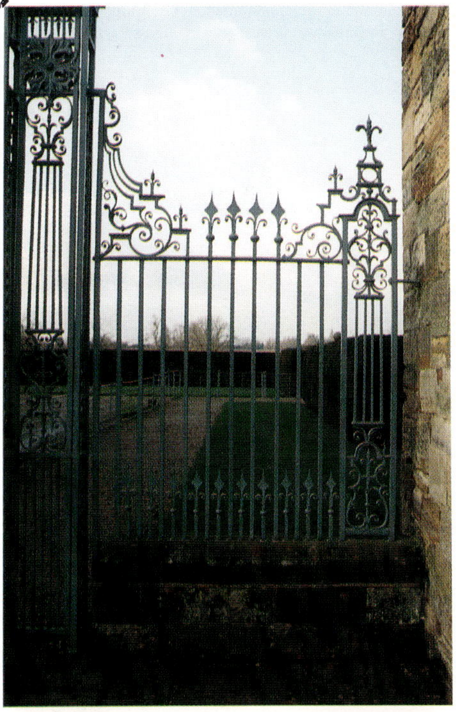

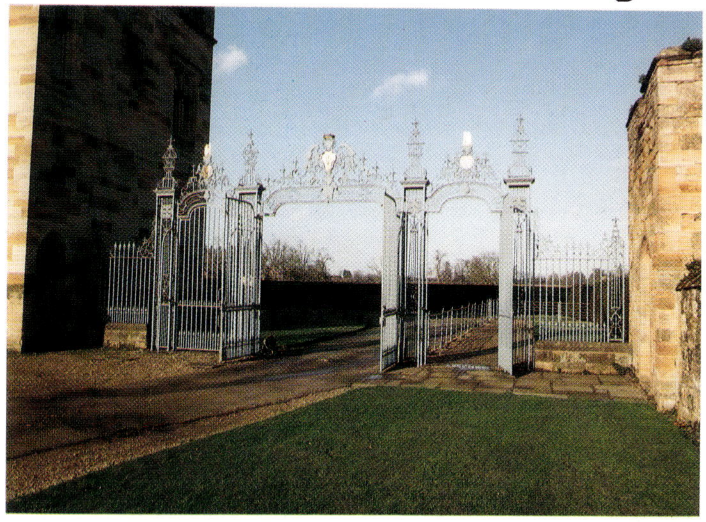

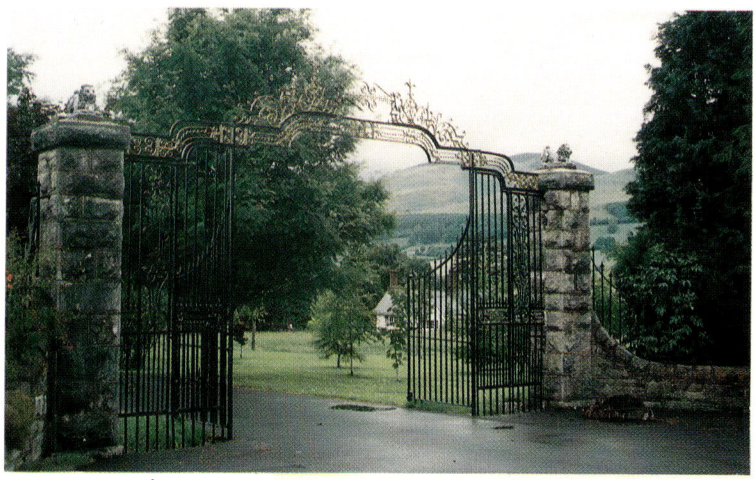

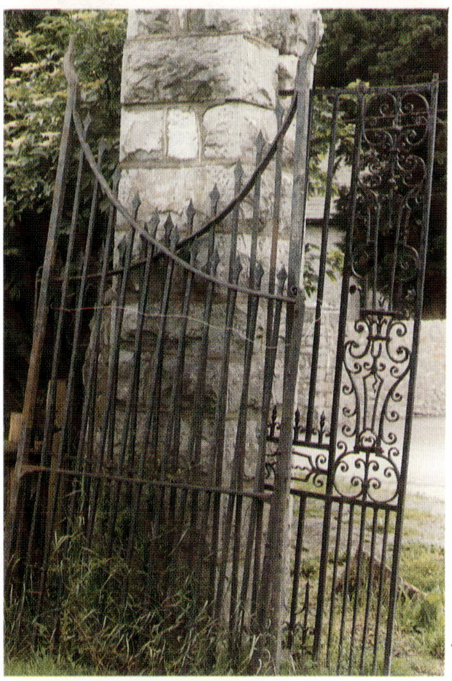

Dignity and Distress...

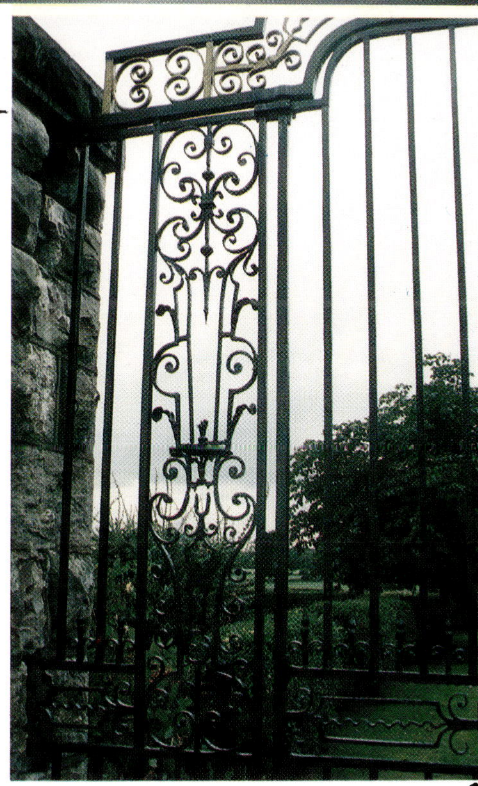

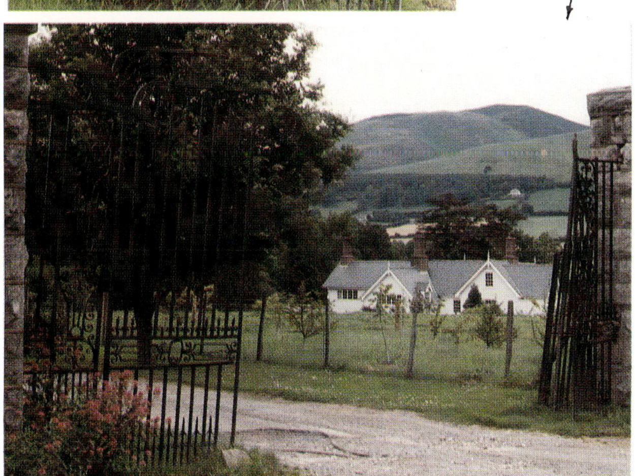

THE SILK MILL GATES

The Derby Silk Mill was the first true factory in this country. It met with success and in 1728 the Lombes obtained the services of Robert Bakewell to make the prestigious gates for the entrance. During the 1920s the gates were restored by the Derby firm of blacksmiths, Taylor Whiting Taylor, who discovered the RB mark upon them. Looking at the gates (Plate 19) it is obvious that Bakewell used the semi-circular profile to gain the maximum lift possible, that is without resorting to the Gothic (see Fig. 9). The upward sweep of the many bead points above the stretcher frame, the stepping stone arrangement of the tables with their water leaf finials, culminating in the high crest of the cross pennants all combine to give pronounced dignity to the whole arrangement. (Officially the cross pennants are termed "Two spears in saltire or, each with a pennant gules").

The addition of the cypher on the central roundel, easily read but obeying all the rules, is a bonus. Perhaps the weakest feature is the mask, but that is a matter of opinion. The side frames with their long straight lines which combine so skilfully with the scrolls, serve to emphasise the upward lift. Some years ago the gates were placed in the Wardwick in Derby at the side of the Public Library. When the Silk Mill was given the task which nothing in the City could better serve, that of the Industrial Museum, the gates were sent back home. The generous gilding on this ironwork is not only a tribute to Bakewell but a welcome sign that the City fathers are conscious of a wonderful heritage.

Plate 19 shows:

Top left	A complete view of the gates and rails,
Top right	A side frame,
Centre left	Details of the overthrow,

whilst at the bottom are two cyphers (Fig. 10). On the left is the original roundel which is now an exhibit in the Bakewell section of the first floor in the building. The remaining picture shows that craftsmanship is still very much alive. It shows the skill of the restorers.

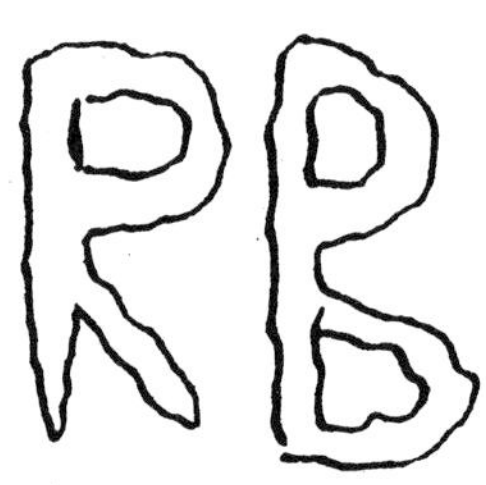

Bakewell's Mark

The Old Silk Mill – now the Industry Museum, Derby.....

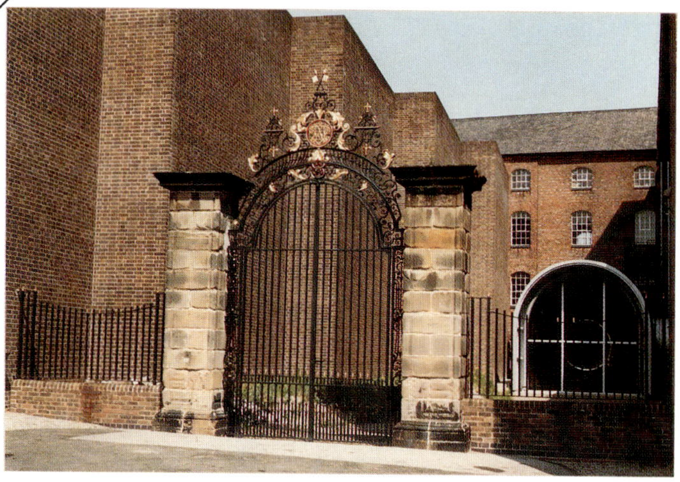

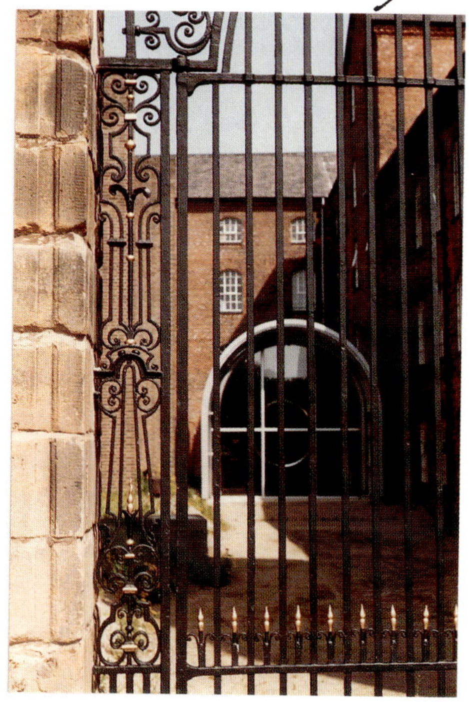

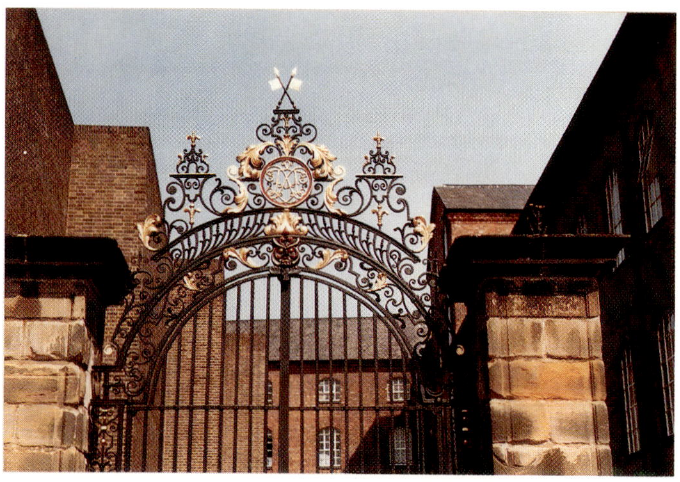

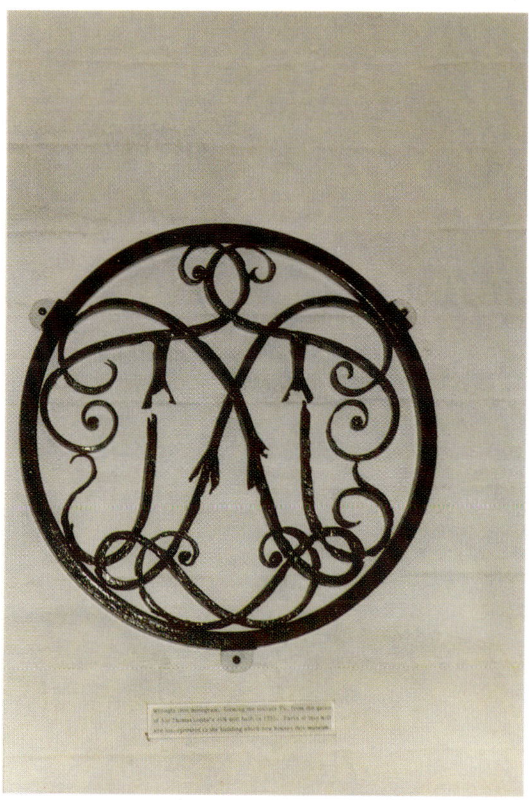

THE 1730s

We now enter the realms of uncertainty, not in attribution but in dating. A number of Bakewell artefacts were made in the 1730s at the time when Francis Smith of Warwick, who worked closely with Bakewell, was very active in the Derby area. There is little doubt that Bakewell would be recommended by many who had experienced or witnessed his work and his clientele would have widened considerably. Other builders than Smith of Warwick would require his services.

LONGFORD HALL, SOUTH DERBYSHIRE

Longford is in the midst of a lovely agricultural area, some eight miles south of Ashbourne. Close by is an interesting Church with an old Preaching Cross. Certain points concerning earlier residents of the hall may have had some bearing on the erection of Bakewell gates there. The blacksmith's first patron was the owner of Melbourne Hall, Thomas Coke. Another branch of the Coke family lived in Norfolk but they purchased Longford as a residence for younger sons. This family connection was probably the reason for further patronage of Robert Bakewell. Though simpler than much of his work, it would also be less expensive - in line with the normal treatment of younger sons in those days. The Bakewell border (Fig. 12) is present, placed vertically in the side frames and in the arc of the cresting (Fig. 8) on the tops of the gates. The dog bars are arrow pointed and the single lock rail is fringed with the usual bead points. Integral with the gates are the crestings; though void of acanthus they are possessed of sweeping, curving scrolls and surmounted by gilded spear points. Very Bakewell!

PARK HALL FARM, CHURCH LEIGH, STAFFORDSHIRE

If the Penshurst gates can be said to be 'Typically Bakewell', here we have another example equally suited to the phrase. We see the inset rectangle, the finger board, the Bakewell border, the single lock rail with the bead point fringe and the spear point dog bars. The main difference lies in the width of the Penshurst screen and the Carriage gates of Park Hall Farm and the layout of the overthrows. At Park Hall Farm the centre of the Overthrow is taken by the fabled figure of a Cockatrice with a dragon for a tail. Although the reason for the inclusion of the animal is unclear, the author believes it to be a kind of evil eye symbol meant as a deterrent to miscreants. On the outside of the stone pillars are grove guards which catch at the clothing of anyone walking too close. If, originally, hedges were grown close to the pillars these guards would collect anyone trying to squeeze between the pillars and the hedge. These gates were beautifully restored at the beginning of the 1980s by a firm of smiths practising in the village of Dove Holes near Buxton. They are excellently maintained by the present owner, Mr. E. J. Knobbs, an octogenarian who keeps a fine herd of pedigree cattle. The opinion has been expressed that these gates once graced the Hall at Trentham in Staffordshire. One thing is certain, they were not made originally for their present situation.

PLATE TWENTY

Longford Hall, Derbyshire

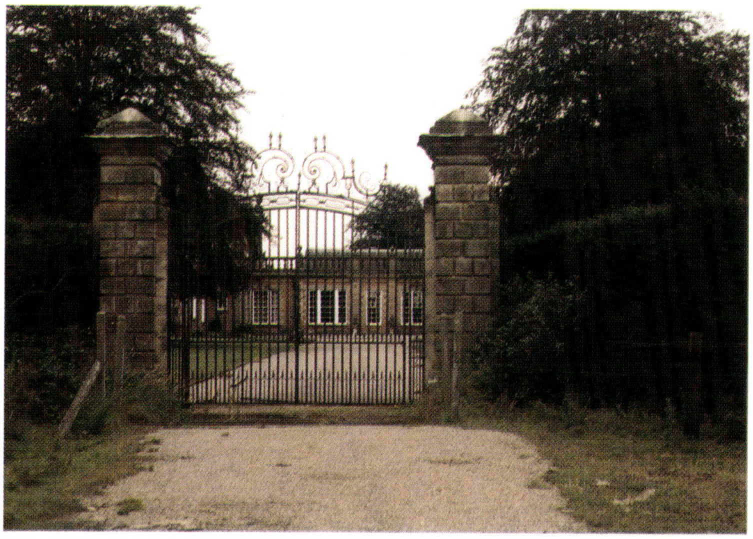

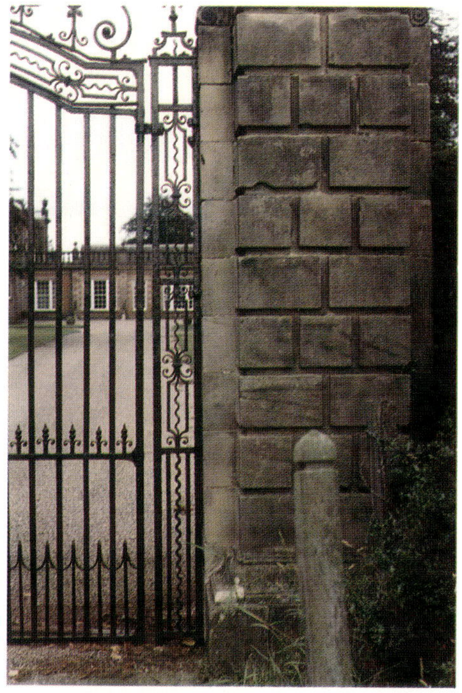

Park Farm, Church Leigh, Staffordshire.

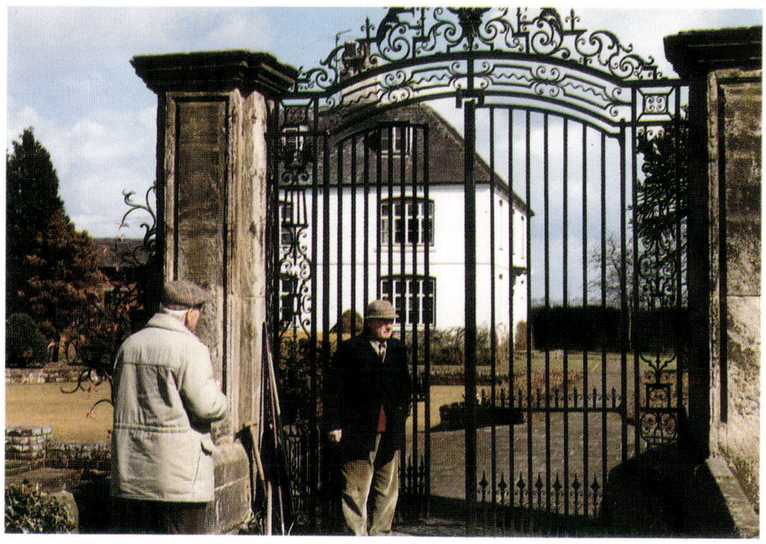

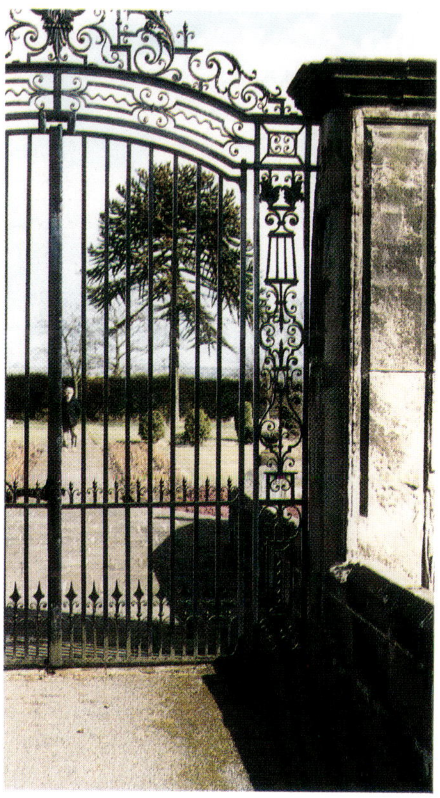

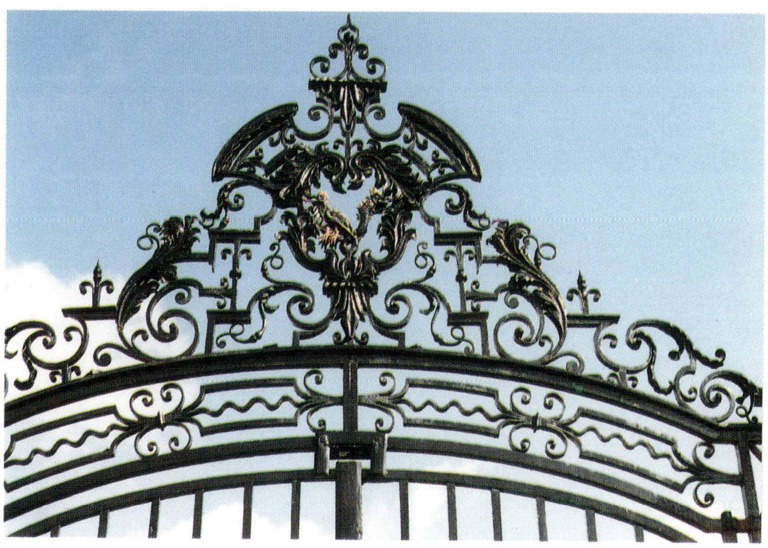

HENLEY HALL, LUDLOW, SHROPSHIRE

Charles Hurt, who married Susannah, a daughter of Sir Richard Arkwright of cotton spinning fame, rebuilt Wirksworth Hall in 1779 and at his death it passed to another Charles Hurt. The second Charles Hurt sold the Hall in 1858 to a Mr. Nicholas Wood. Early in this Century the family of Nicholas Wood left the district to take up residence at Henley Hall, just outside Ludlow. It was fortunate that the Bakewell gates and rails were taken with them. Plate 21 shows the present state of this lovely ironwork. At Top left is seen a long view of the ironwork, which is very different in height from Bakewell's usual style, but handsome nonetheless. Here is an example of crested gates, gates which have the bulk of their decoration built on the top rail as distinct from a separate overthrow. (See Fig. 8). When these gates oscillate the decoration swings with them.

The main gates (Bottom left) are taller versions and only slightly more elaborate in the cresting than the wicket gates (Centre left). The flat standards on the same picture are diminutives of the flat pilasters (Top right). The dog bars are unusually tall and are spear pointed; the spear points being repeated as a fringe on the pallisades.

The inset rectangle and the finger board arrangement make up a large part of the standards and pilasters which are crowned (Fig. 5) by an open bar vase form and finally topped by a naturalistic finial.

The picture (Bottom right) shows a terminal bracket (Fig. 17) complete with mobile acanthus. On the cresting of the gates it is possible to discern three small acanthus motifs. These are exactly like those on the terminal bracket at the Worcester Guildhall. (Plate 15).

The appearance of the Henley Hall gates would be much improved if the leaves and flowers were to be gilded.

OSMASTON HALL, DERBY

The Hall was built, apparently in the manner of Francis Smith of Warwick, who so frequently relied on the work of Robert Bakewell to add a final flourish to his buildings. It was erected about 1700. Craven and Stanley, in their first volume of *The Derbyshire Country House* tell us that "Vanished Bakewell Gates opened on to a Park". The Hall was demolished in 1938, and with it, presumably the gates to the Park. The Wilmot Horton family who owned the Hall for many years left to take up residence at Catton Hall, near Burton-upon-Trent early in the nineteenth century.

Mr. D. W. H. Nielson, a descendant of the family allowed the author to have this drawing copied. It is attributable to Bakewell and is dated 1750 on the reverse.

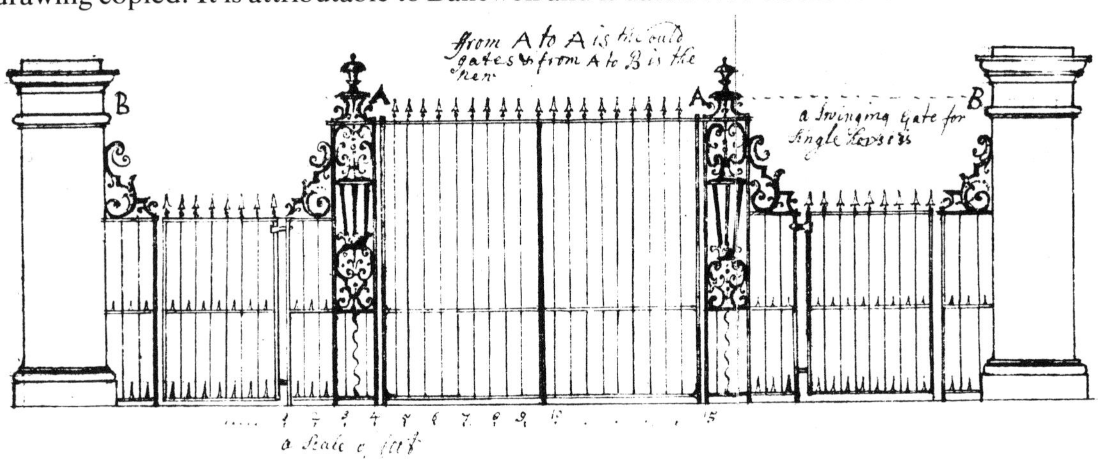

Courtesy Mr. D. W. H. Nielson and the Derbyshire County Library.

Henley Hall, Ludlow, Shropshire. Gates ex Wirksworth, Derbys.

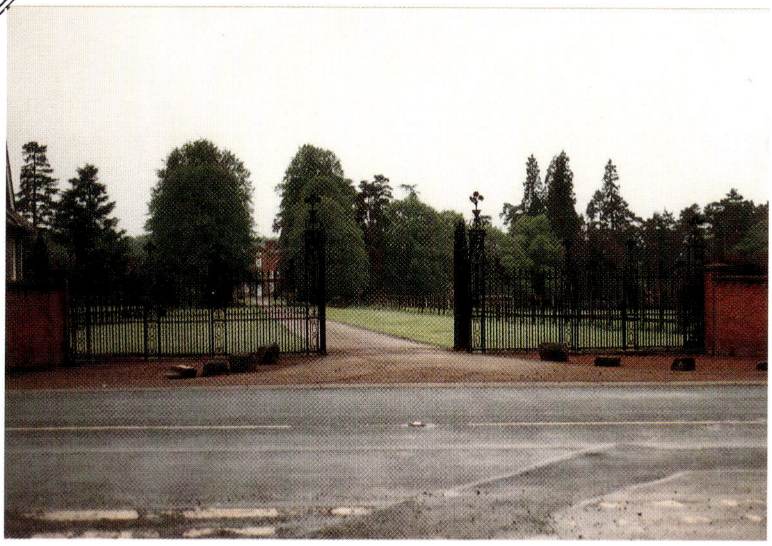

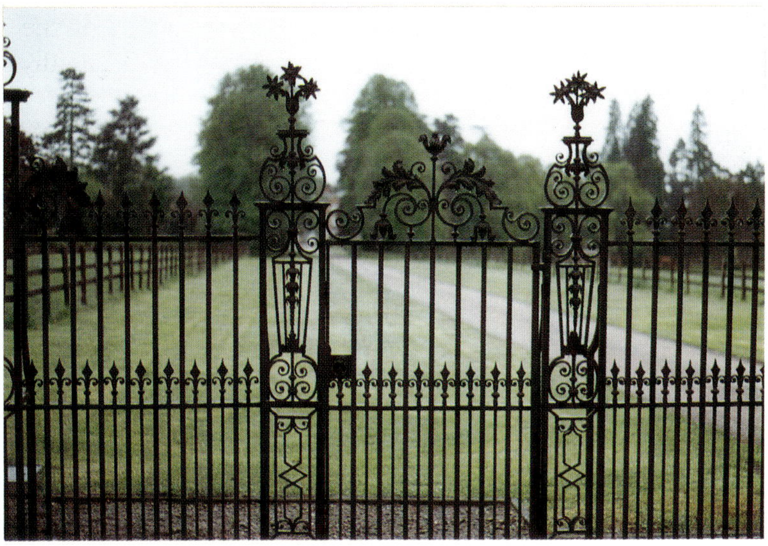

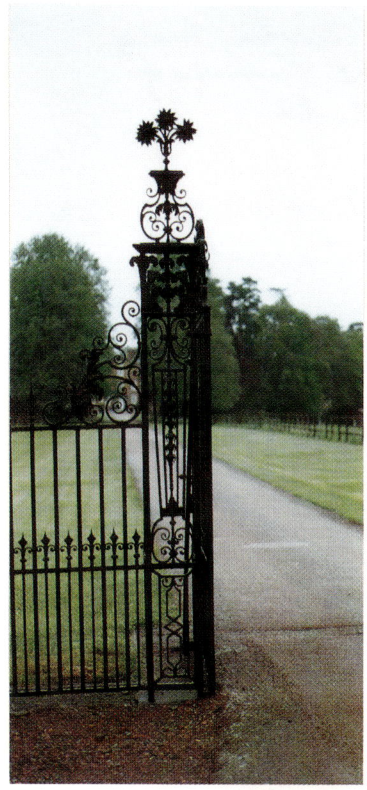

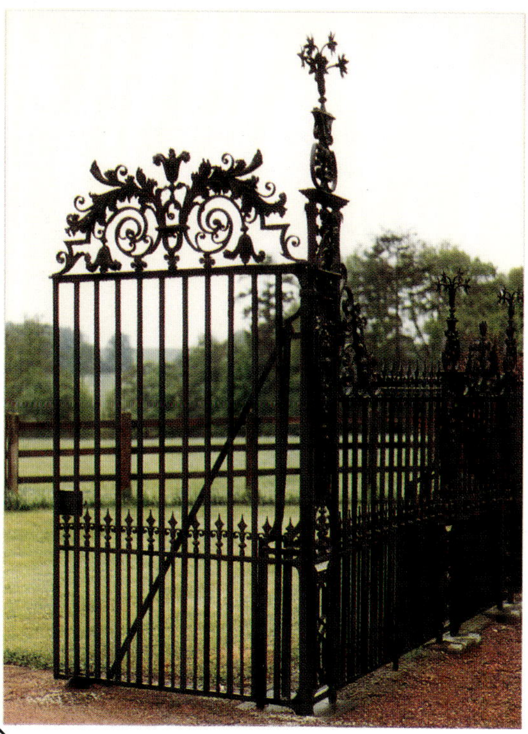

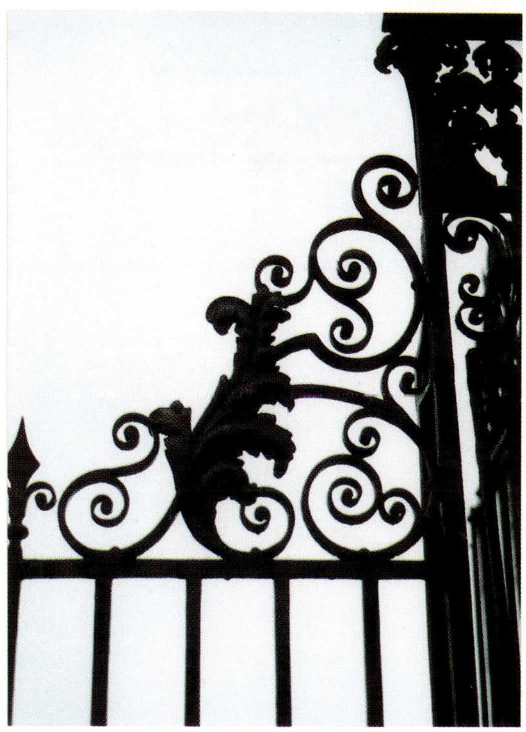

NEWNHAM PADDOX, NEAR MONK'S KIRBY, WARWICKSHIRE

When Saint Mary's Gate House, near Derby, was demolished the Bakewell ironwork there was sold. The main gates are now (since 1958) at the West End of Derby Cathedral, the perron ironwork is at Barton Hall, Church Broughton but one gate, which a drawing of appears below, is no longer extant. The nearest duplicates of this gate appear at Newnham Paddox, on the Denbigh Estate. There are differing ideas concerning the origin of this screen (see opposite). It has been written that the work was originally ordered by a French nobleman who defected; that it was made for Bewick, near Shrewsbury by the Brothers Davies of Croes Foel, Bersham near Wrexham; and that when it was transferred to its present site Robert Bakewell was asked to make two wicket gates in order to complete the required span. Some of this may be true, some may be fable. However, one can accept that the wicket gates are Bakewell's on stylistic evidence. There are pointers, too, that he designed all or part of the remainder of the work. The roundel on the huge side frames, the naturalistic finials atop the pilaster, the scrolly pattern reminiscent of the Silk Mill gates, the lock frame with the inset rectangle and the wavy line, and the side frames which repeat both this motif and the lyre form speak eloquently of the Uttoxeter-born genius.

Messrs. Ayrton and Silcock in their treatise *Decorative Wrought Ironwork (1929)* Mention that the Davieses and Bakewell worked together in the early stages of their careers. There is no record of this but it is commonsense to assume that if the order books of either of the two firms was overloaded they would seek assistance elsewhere. Bearing in mind that Bakewell had his own staff this would not be a frequent occurrence but it would not be unreasonable to suppose that on a large work such as this, Leeswood Hall screen or Eaton Hall gates, the two firms could well come together in mutual assistance.

It is also possible that Bakewell was the designer of these screens as his reputation was built as much on his innate sense of proportion as on his undoubted skill as a craftsman. He has been criticised by some commentators for having plain gates and confining his major decoration to the overthrow, side frames and pilasters. This, of course, misses the point. His gates stand tall, uncluttered and dignified, largely unbroken by horizontal decor.

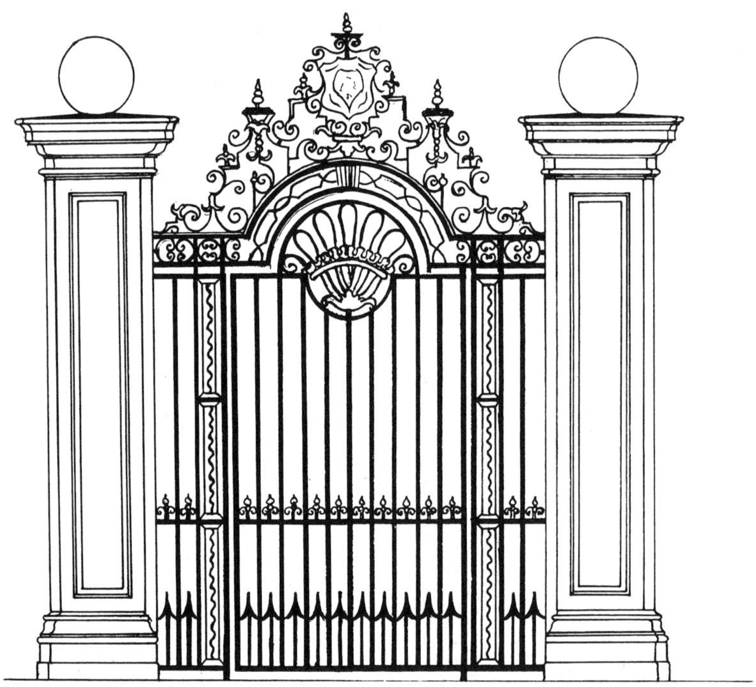

St. Mary's Gate House, Derby. Single Gate. c. 1725. (Page 62).

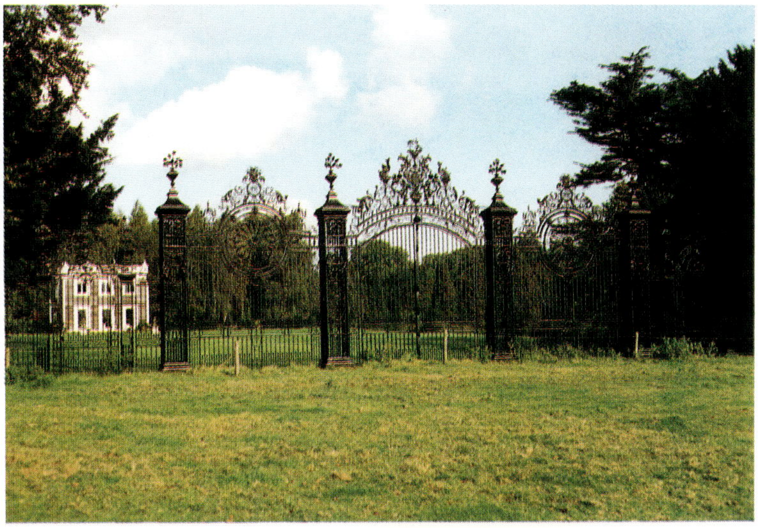

Newnham Paddox,
nr. Monks Kirby,
Warwickshire.
The Great Screen.
Wicket Gates by
Robert Bakewell....

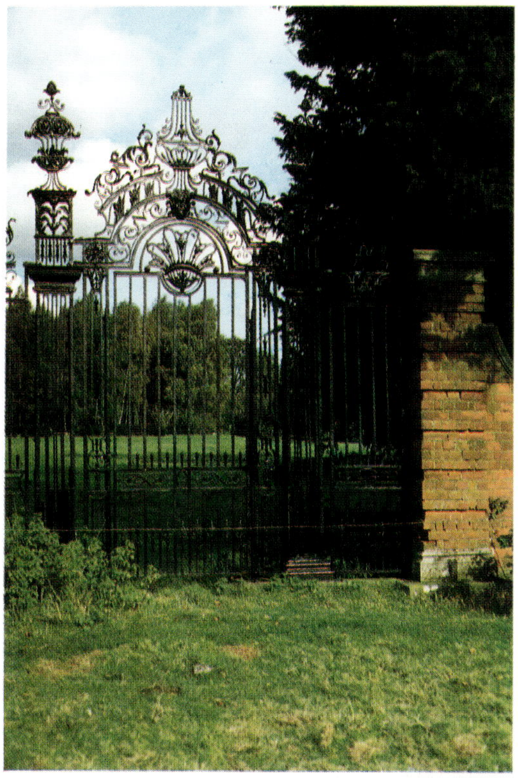

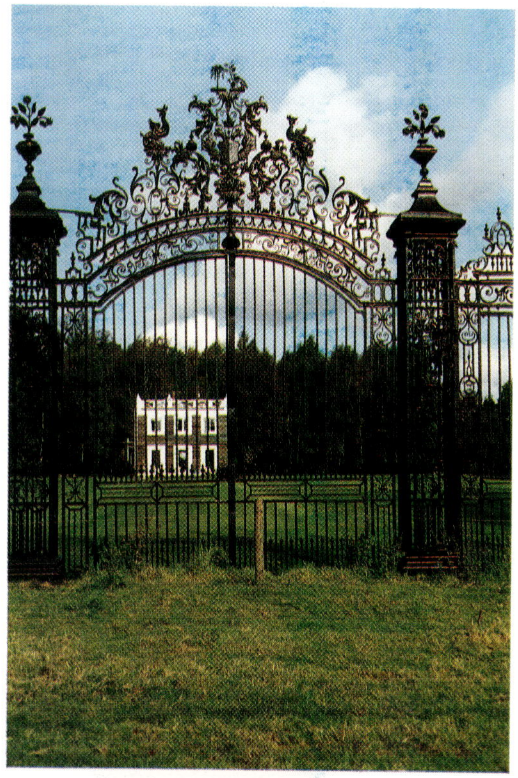

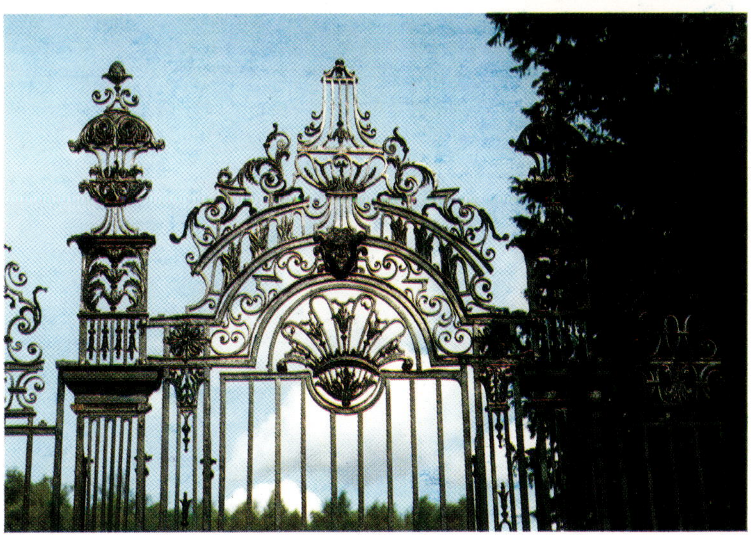

THE CARRIAGE GATES, ETWALL, NEAR DERBY

Etwall Hall, the home in the sixteenth century of Sir John Port who was the founder of Repton School, was demolished in 1954 in order to make way for the John Port Comprehensive School. In the seventeen thirties the Hall was the residence of Rowland Cotton and his wife née Mary Sleigh. They were responsible for ordering the Bakewell gates. Having withstood over two hundred years of rain, snow, tempest and flood the gates were removed in 1955 by the County Council. Years later they were found rusting in a scrapyard in Clay Cross, near Chesterfield. A number of committed Etwall residents were responsible for recovering them and having them stored. The Hawk and Buckle crest and the arms of Cotton with Sleigh were rescued from a person generously described as an amateur collector. An amazing effort by the residents of Etwall, surrounding areas and even further afield contributed to the cost of restoration. The erection was carried out free of charge by the Civil Engineering firm of Roger Bullivant. The Committee was chaired by Lady Loudon, a direct descendant of the Port Family and she and the other members chose Mr. Peter Walker of the Chapel Cottage Forge, Anslow, Burton-upon-Trent, to carry out the restoration. The choice was influenced by the recent successes he had had in National Competitions. He was said to be able to work using the same methods as Bakewell himself.

Plate 23 shows:	Top left	The gates as re-erected in 1985.
	Top right	The tribute to the restorer.
	Centre right	The original entrance to the Almshouse quadrangle.
	Bottom right	The illustrations used by the restorer.
	Bottom left	New water leaves just attached, in the workshop.

The original entrance to the Almshouses consisted of two narrow gates giving about five feet of clearance from brick wall to brick wall. The new distance from wall to wall is twenty two feet six inches. The overthrow of these gates, though not Bakewell's largest by any means, shows the liveliness of the acanthus and the natural appearance of Mr. Walker's tulips. The pilaster finials are equipped with cloths of estate in blue and gold and the central cartouche with the arms of Cotton with Sleigh.

The Restored Carriage Gates Fronting the John Port Almshouses, Etwall, Derby.

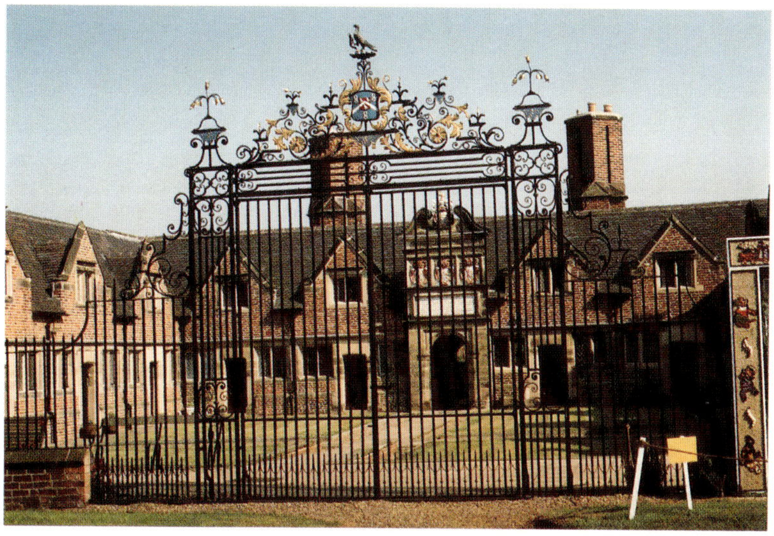

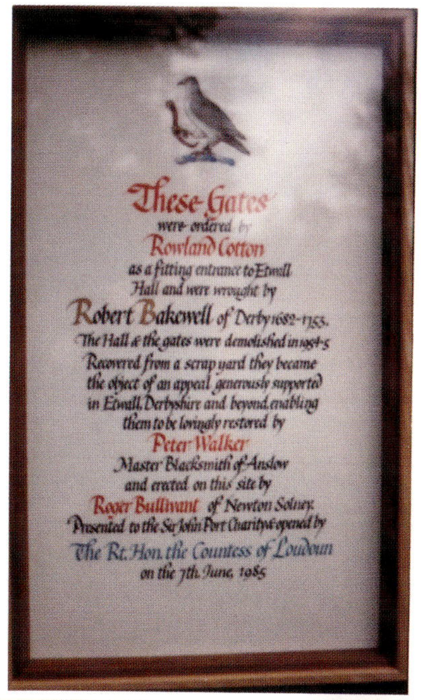

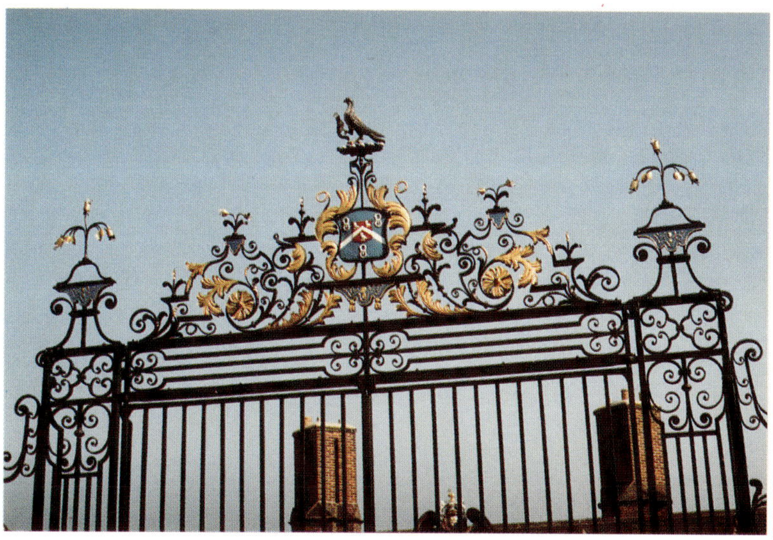

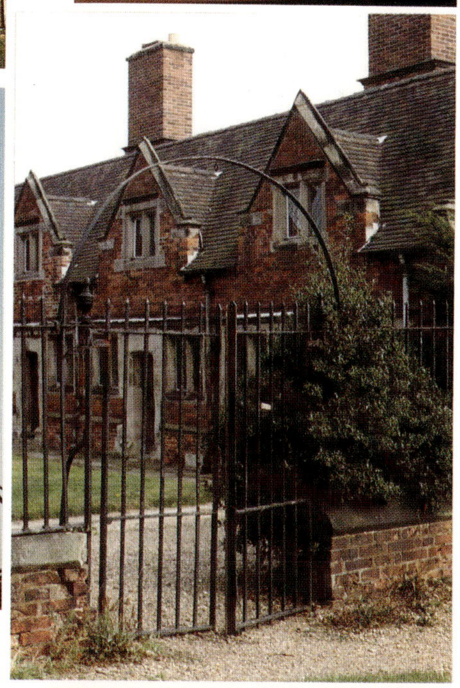

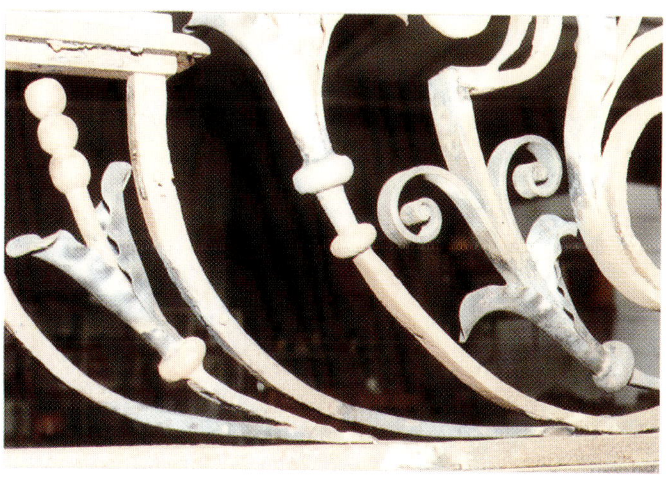

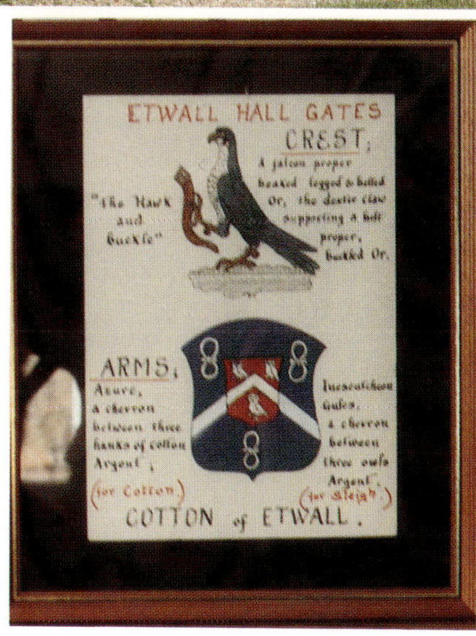

OSBORNE HOUSE, SAINT MARY'S GATE, DERBY

In 1725-27 Edward Osborne who had been the most generous non-titled contributor to the rebuilding of All Saints' Parish Church (now the Cathedral) in Derby had a house built about one third the way down Saint Mary's Gate. It is frequently referred to as Saint Mary's Gate House. In charge of the building of this house was Francis Smith of Warwick and responsible for a great deal of ironwork of outstanding quality was Robert Bakewell. Several owners came after the Osbornes including the Evans family, mill owners of Darley Abbey and, last of all, the Baptist Congregation who adapted the whole of the inside as a Chapel. In 1937, due to increased maintenance costs and a falling attendance, the Chapel was sold to a large motoring organisation. Unforgivably, this commercial concern demolished all but one wing of the place. Fortunately the ironwork, in the main, was saved for posterity. The noble gates were rescued, purchased and restored with the assistance of the Friends of Derby Cathedral and sited at the West End of the Cathedral. Even though they are built around a perfectly flat stretcher frame they are possessed of undeniable dignity. The pilasters are of rectangular section and the main decorative features are in an inset panel which is seen when looking either to the front or back. The side decoration is very simple.

The Bakewell border is present as is the wavy line and there are three versions of the inset rectangle. There are also two innovations on this ironwork; one is the pendant fringe on the gates and pallisades, with its leaf endings (Fig. 7) and the upright fringe found only on the pallisades and the Chambers' monument inside the Cathedral. (Fig 13 and Plate 28). The overthrow is most attractive, fashioned around a cartouche bearing the Osborne Arms. The rather skilful way in which the gates are painted and gilded turns work that is already special into something extra special. The use of gilding to enhance wrought iron has already been referred to, and the Osborne gates are, perhaps the most suitably gilded of all the work of Robert Bakewell. These gates are so neat, there is so much detail and so much busy-ness that it is impossible to over-praise them.

At the entrance to Osborne House, on the platform or Perron was some delightful ironwork featuring some balustrading and the Bakewell Roundel that would not disgrace a Church building as the frame of a rose window. This work now graces a feature of Barton Hall, Church Broughton, near Sudbury. (Plate 25 and Fig. 15).

A charming single gate which used to be at Osborne House was praised for its design which featured an open cockleshell feature set in its semi-circular profile. This gate is no longer extant but a sound idea of its form can be gained by viewing Plate 22, Newnham Paddox, and the drawing on Page 58.

In 1731 Bakewell was paid £50. 19s. 8d. plus 1s. 0d. for Ale in the payment for a gate in the Churchyard of All Saints', Derby.

'All Saints' Cathedral, Derby.
The West End Gates. c1725

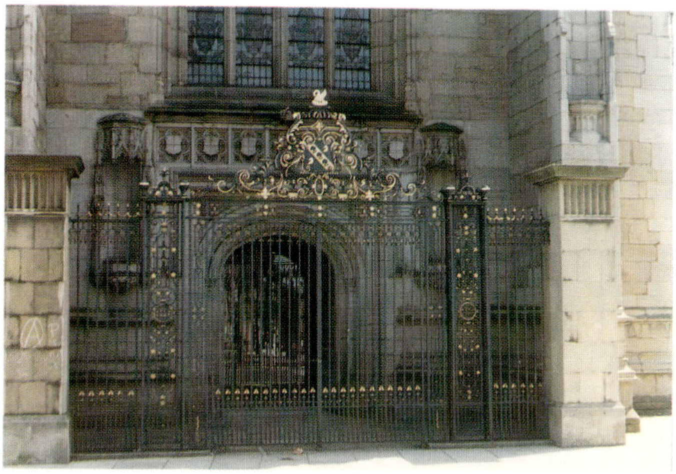

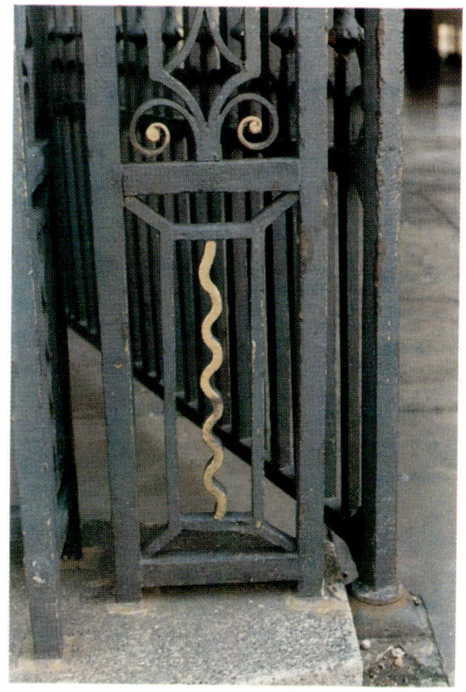

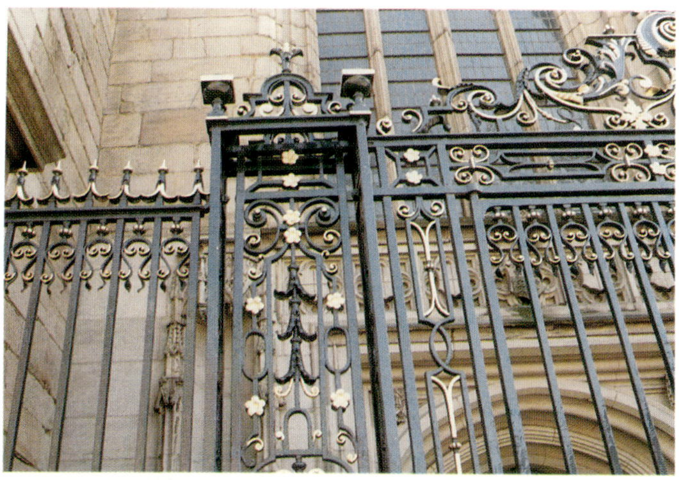

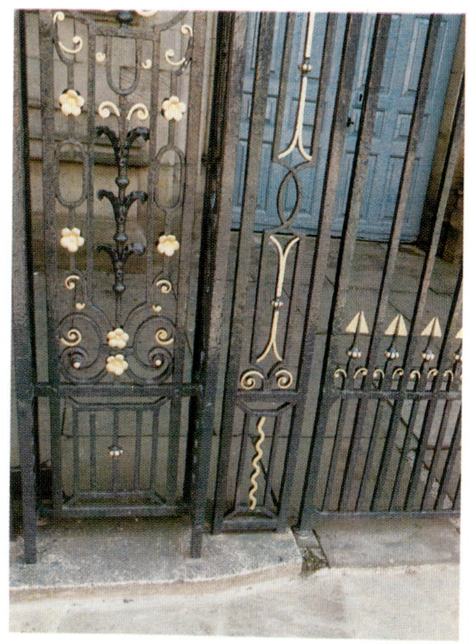

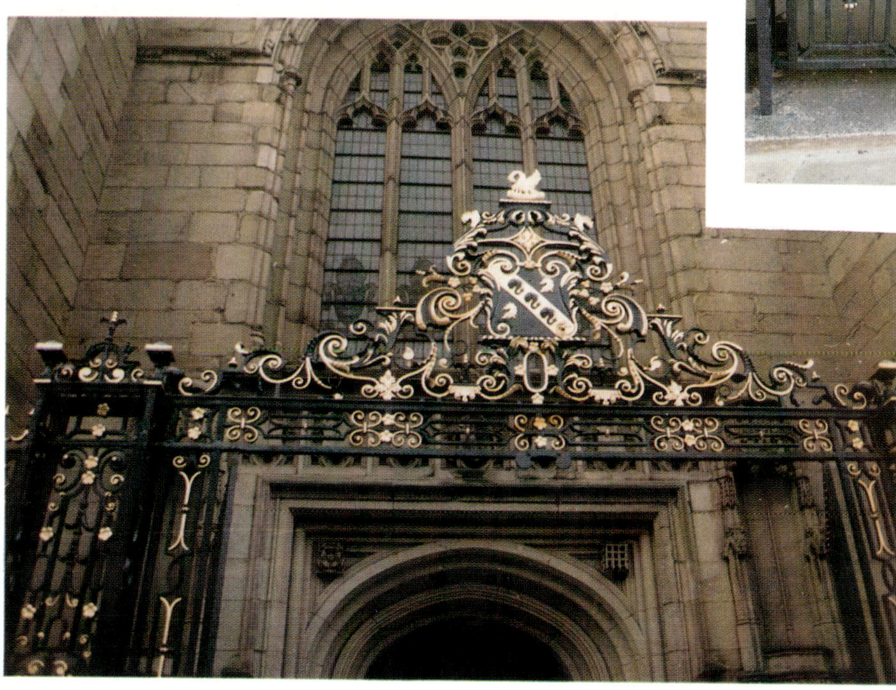

Made by Bakewell for Saint Mary's Gate House.

BARTON HALL, CHURCH BROUGHTON, NEAR SUDBURY, DERBYSHIRE —

The Bakewell ironwork from the Osborne or Saint Mary's Gate House was originally taken into store when the house was demolished. At different times it has become separated so that the gates are at the West End of the Cathedral in Derby (Plate 24) and the perron ironwork is now at Barton Hall, the home of Mr. and Mrs. J. W. Pratt. Top left, Plate 25, can be seen the platform, or perron, the steps and the impressive ironwork, which is kept in quite superb condition. Viewing the roundel (Centre left), it is possible to appreciate why, on several occasions, the author has praised gilding for its value in emphasising good craftsmanship. In this case the gilding has highlighted the repoussé in a way that light and shade alone could not have achieved in black sheet iron.

Considering the roundel as a motif in itself, it is possible to appreciate that Bakewell was not merely a good blacksmith, not merely an artist, but that he had an obsession for detail. After all there must be many, many Rose Windows in Churches throughout the land that cannot compare with this roundel.

The balusters are typically Bakewell, the lyre form and the presence of the spur are convincing.

THE BAKEWELL RB MARK

Mr. Alan Knight a leading artist - blacksmith and authority on the history of wrought ironwork was given the task of restoring the gates which front the Almshouses at Preston-on-Weald, Telford, Shropshire. Whilst replacing parts too weathered to remain, he discovered the RB mark on one of the cross rails or transoms. This portion of it (Top right, plate 25) he gave to Mr. E. J. Saunders R.I.B.A. who in turn generously passed it on to the Derby Museum Service.

The signature is punched into the metal. This process, it is believed, increases the density of the metal in such a way that wear and tear affect the punched part more slowly. The actual signature, then, in spite of rust all around it, tends to stand proud. Unfortunately the specimen in the picture on Plate 25 is not photogenic and the ellipse drawn around the signature serves only to locate and not to clarify the subject. The RB is near the upper end of the major axis. It has been said that no item of ironwork can be 100% attributable to Bakewell without this signature but if that were true we could guarantee but little and the evidence of one's own eyes would be set at nought.

DUNHAM MASSEY, CHESHIRE

The author has it on good authority that amongst the Dunham Massey papers is an account book dated 1732 - 1750 which contains the following entry:

"Mr Bakewell, the Iron Man at the New Building £19. 6. 0".

On the page opposite, Bottom right, is a small garden gate having arrow pointed dog bars and water leaves. Next to it is the picture of guard rails near the pond. Both these items are at Dunham. The statement in the account book is so bald and so bare as to yield no clue whatever as to the work for which the payment was made.

Barton Hall, Church Broughton.

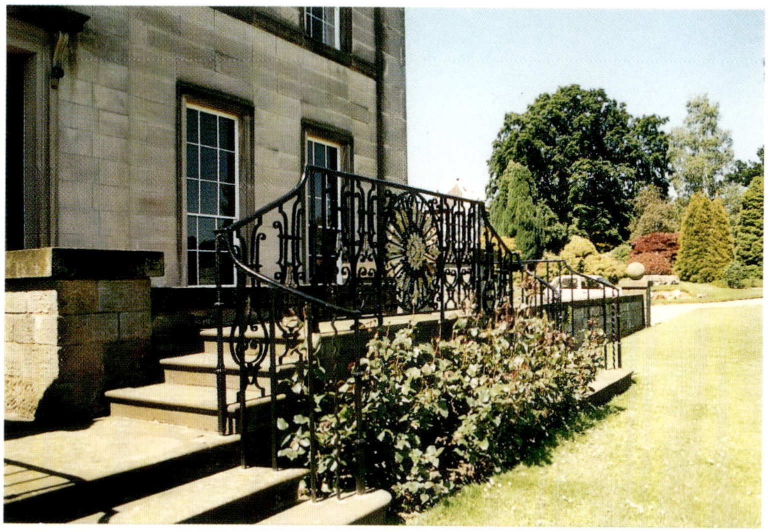

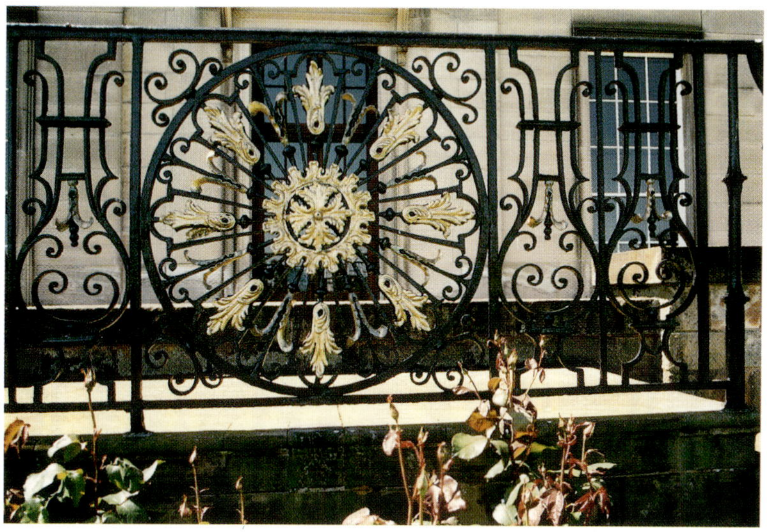

Dunham Massey, Cheshire.

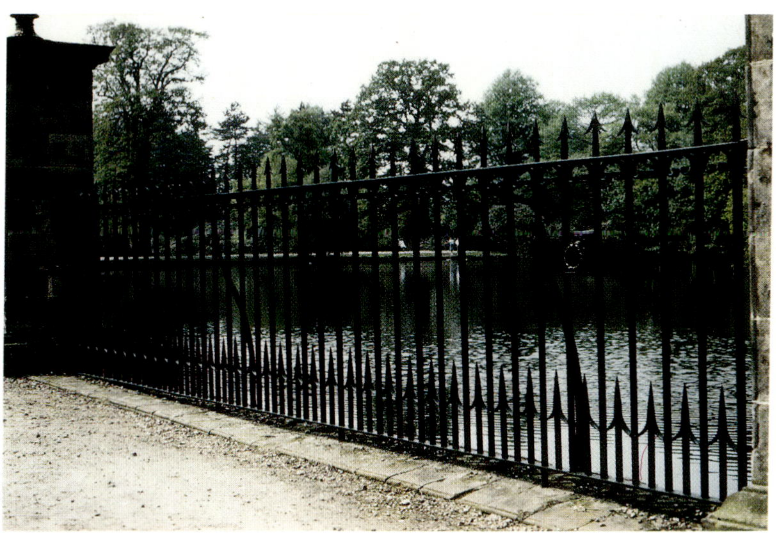

The Industry Museum, Full Street, Derby.

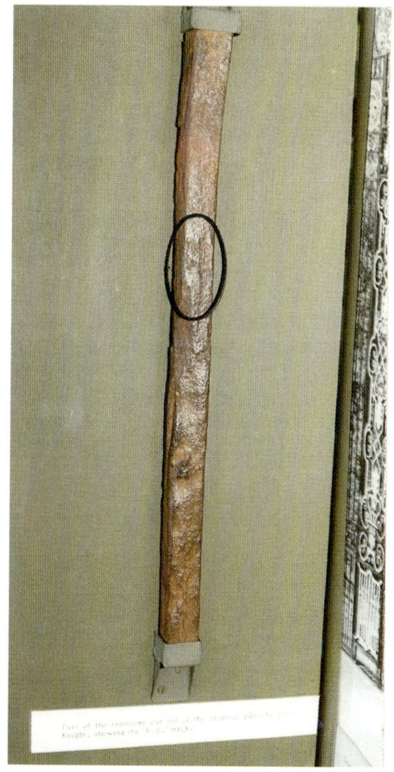

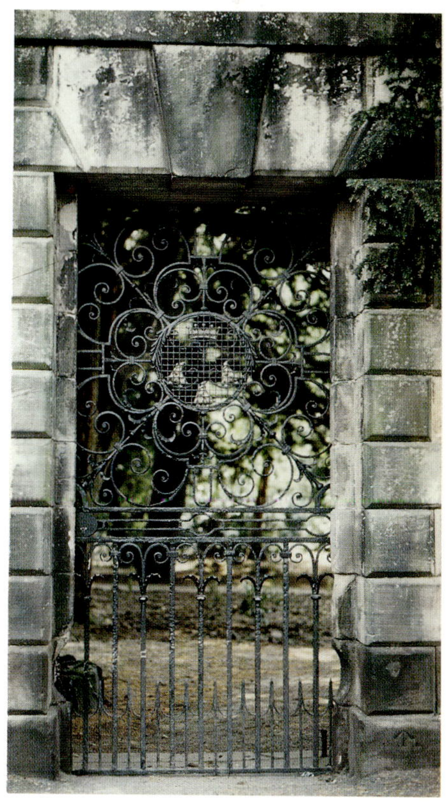

WEST GATE HOUSE, BIRD STREET, LICHFIELD

The author has been unable to trace either the builder of the house or the maker of the ironwork. There are, however, one or two points worthy of note. The house is obviously Georgian even though it is small by comparison with most of the buildings attributed to Francis Smith of Warwick. The ironwork has one or two claims to partly satisfy Robert Bakewell enthusiasts. In the picture (Plate 26, Top right) for instance, one can see two wavy lines at the base of the pilaster making a thunder and lightning panel. (Fig. 4 and Plate 4). On the top of the pilaster is the open bar vase form surmounted by the kind of leaf finial seen on other work such as at Foremark and at Okeover. The gates are equipped with a single lock rail with the now familiar bead points. The dog bars are arrow pointed. The pallisades are probably later than the gates. The house itself and the ironwork in its own right are both Grade II listed.

THE ABBEY HOUSE, SHREWSBURY

On Plate 27 the gates at Preston-on-Weald are pictured. These are definitely Bakewell gates. It has been said, not without some justification, that he who made the Preston gates made those at the Abbey House, also. This is due in large measure to the similarity of the scroll work in the stretcher frames, the general mood of the overthrow and the similarity of the panels in the gates.

WENTWORTH WOODHOUSE, NEAR ROTHERHAM, YORKSHIRE

Shown on page viii is the West front of Wentworth Woodhouse. The house is a conglomerate, a mixture of a bright, lively West front with a staid, dull by comparison, East front. Both were built for Thomas Wentworth, Lord Malton, who later became the first Marquess of Rockingham. The property was left to him in 1723 and he built assiduously from 1725 until 1735. He lived there for twenty years, dying in 1750. Facing Plate 9 there is a memorandum, dated 1716, of an agreement between Bakewell and the Earl of Strafford. This Earl was the uncle of Thomas Wentworth's father. He also called himself Thomas Wentworth having changed his surname from 'Watson'.

The balusters shown in the photograph are 90% identical with those at the Homestead (Plate 28). They are of lyre form, they have the Bakewell spur and differ only in that at the bottom of the belly of the lyre, Wentworth have a bead point, whilst at the Homestead they have a C scroll. The terminal panel at the base of the steps in both cases as well as at Barton Hall (Plate 25) has its wavy line to help to convince us that they all came from Bakewell's shop.

West Gate House, Bird Street, Lichfield, Staffordshire.

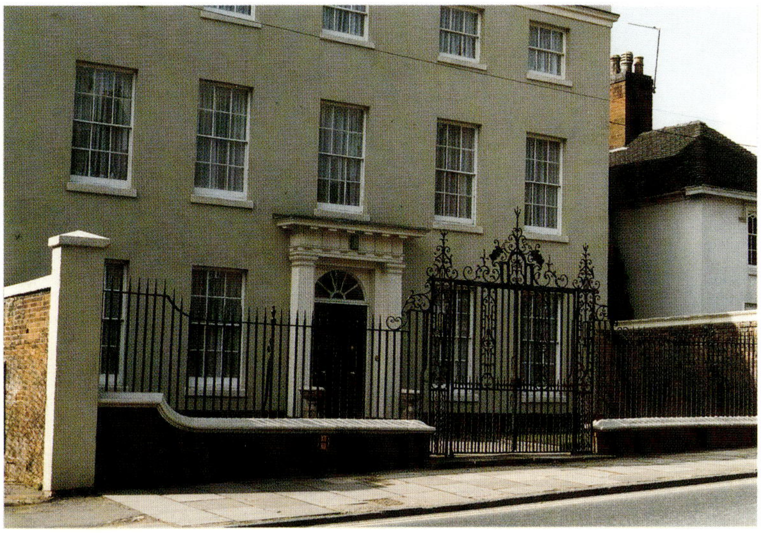

Both House and Gates are 'Grade 2' listed.....

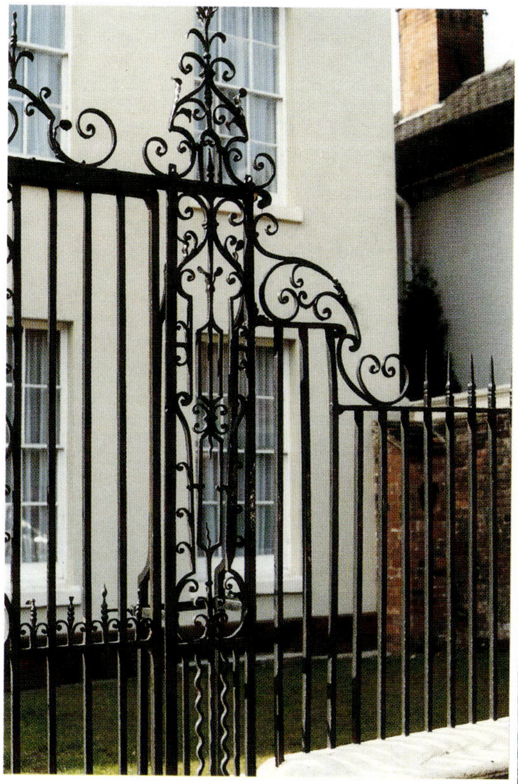

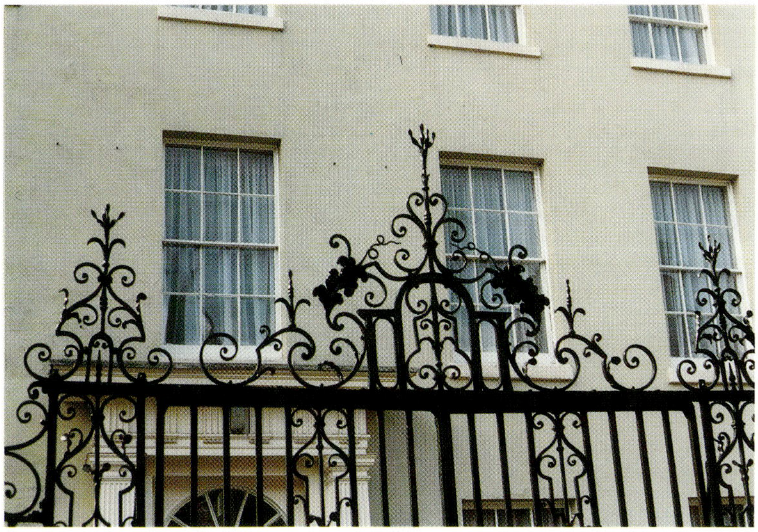

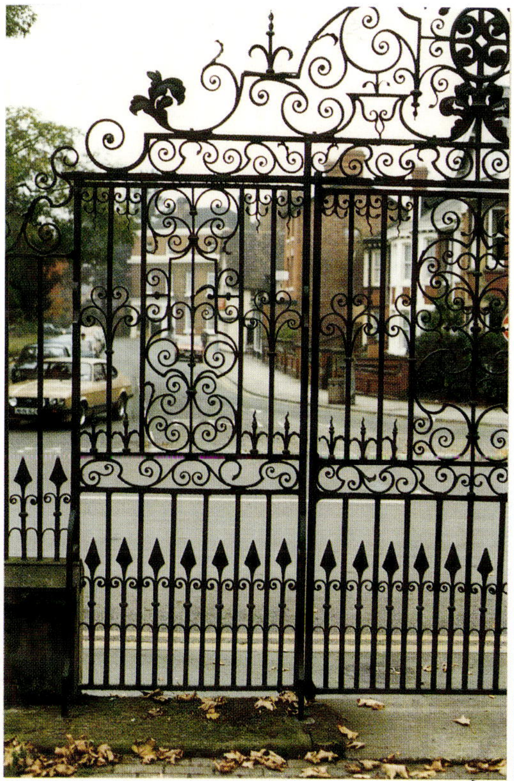

The Abbey House, Shrewsbury...

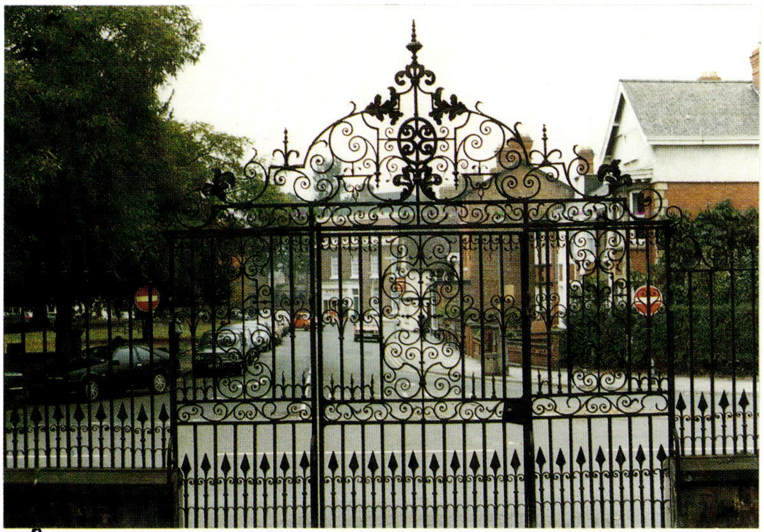

PRESTON-ON-WEALD, TELFORD, SHROPSHIRE

Alan Knight, who lives in Hanbury, Worcestershire was, on one occasion responsible for restoring the gates which stand in front of the Almshouses at Preston-on-Weald. He found it necessary to replace certain items, one of which was a transom rail. (Plate 25). On this he found the undisputable Bakewell trademark, the RB signature.

Although the gates can hardly be thought to be typically Bakewell it must be remembered that certain of the blacksmith's patrons would have ideas and minds of their own and possibly their own designs.

Plate 27 shows the gate quite clearly. The dog bars are spear pointed and the pendant fringe below the locking frame is reminiscent of the one on the Chesshyre gate (Plate 13) and one on the Osborne gates (Plate 24). The pallisades outside the gated area are of a much later date and were cast at Coalbrookdale. Architecturally the Almshouses are very interesting in their own right. In the rear is an old pump still very much in form.

HIGH GREEN, CANNOCK, STAFFORDSHIRE

In High Green stands a most attractive Georgian house which until recently was used as the main Council Offices. It has recently been replaced by a large purpose-built suite of offices some few hundred yards away. The gates are not documented, however, so that doubt has been expressed as to their origins. We can only be guided by style, therefore, and to some extent by the age of the house.

The dog bars are arrow pointed, a Bakewell feature. There is, too, a lock bar rather than a lock frame, a second Bakewell feature, and in order to convince the Bakewell enthusiasts, there is a bead point fringe. Turning to the side frame we see the Bakewell border and see it also in the semi-circular stretcher frame beneath the overthrow.

The overthrow is beautifully scrolled and the acanthus shows the Bakewell mobility. The central cartouche is empty of everything but its frame. Had it been complete we would have known more concerning its origins.

A comparison of the pilasters on the Cannock ironwork with the side frames of the Park Hall Farm (Plate 20) and Penshurst (Plate 18) will show how frequently certain motifs were used by Bakewell. In all three we have the inset rectangle with the wavy line, the lyre form and the finger board construction, albeit in varying proportions. If Bakewell did not make the gates he at least designed them.

It is pleasing to note that these gates have been selectively gilded; it points to an appreciation of good craftsmanship.

PLATE TWENTY SEVEN

The Almshouses, Preston on Weald, Shropshire...

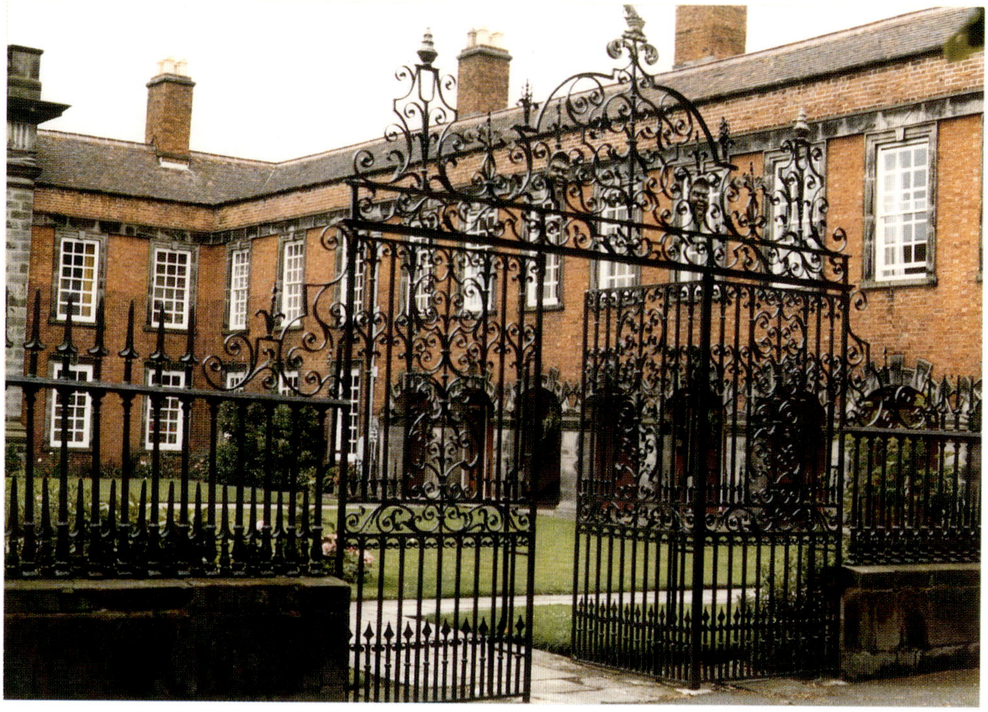

High Green, Cannock, Staffordshire, Former Council Offices.

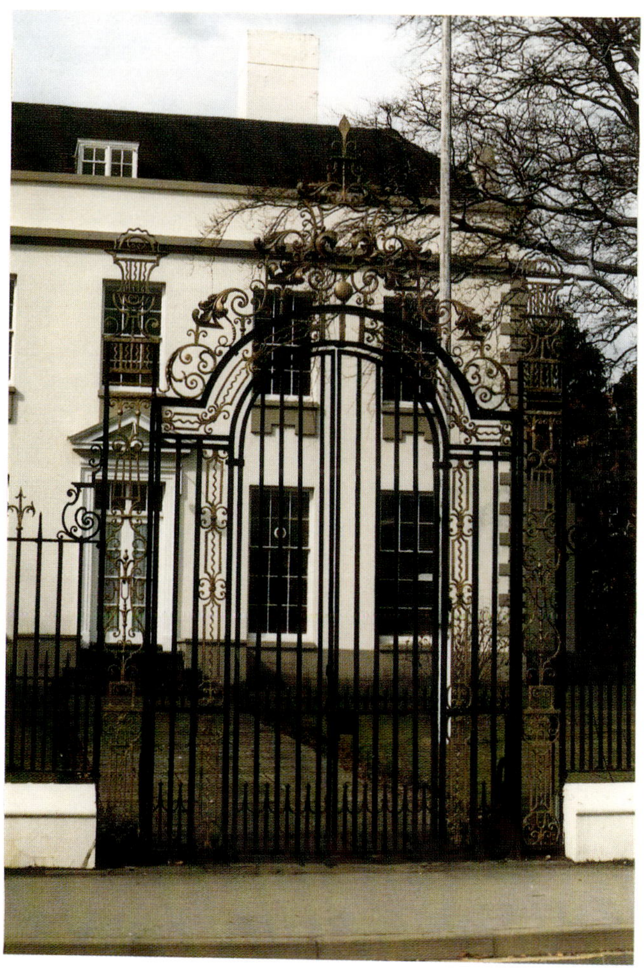

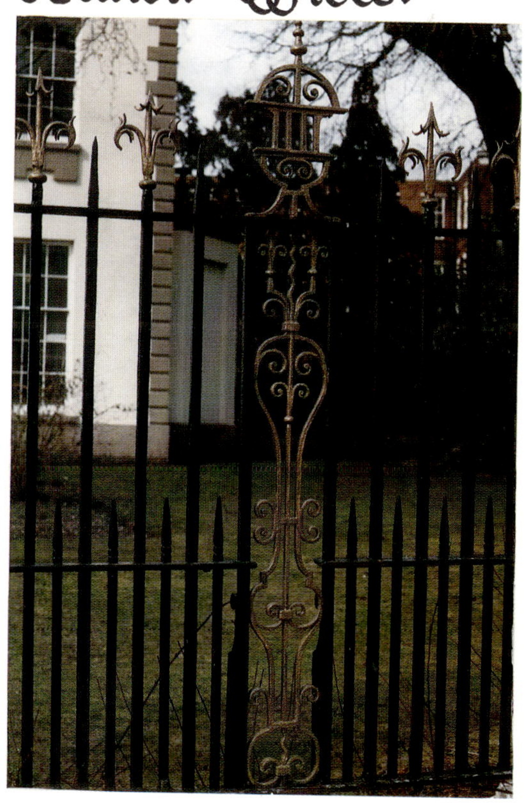

PERRON IRONWORK AT THE HOMESTEAD, SPONDON, DERBY

This lovely house is owned by British Celanese and is host to visitors to their factories in the area. As can be seen at the top of Plate 28 it is fronted by some splendid ironwork, which it is said, came from Darley Hall, Darley Abbey, on the North East of Derby.

According to Craven and Stanley in *The Derbyshire Country House, Volume 1,* Darley Hall was built in 1728 by Francis Smith of Warwick. "The entrance was by an attractive stairway, with a Gibbs surround and a wrought iron handrail by Bakewell". It was demolished in 1960.

The platform by the entrance is known as a 'perron' and 'perron ironwork' must therefore refer to such balustrades as we see at the Homestead. Fig. 3 'Bakewell Balustrades' will show how close this example is to other work by him. The lyre form is obvious as is the spur which, as already indicated is almost a signature of his. Across the top four balusters on each side of the perron is the Bakewell border, whilst in the terminal arrangement at the foot of the stairway is the wavy line. The turned finials, too, compare accurately with those on the garden arbour at Melbourne.

CALKE CHURCH, CALKE ABBEY, TICKNALL, DERBY

In the Abbey grounds stands the Church where there is a monument to Sir John Harpur who died in 1741 and his wife Catherine (née Crewe) who died in 1745. In her will she requested her son to erect a plain monument to his father. In the event he paid Sir Henry Cheere £200 to carve the work in marble, bearing the busts of both his parents. Simple, very ordinary rails were set up by Bakewell in 1746. An extract from the Harpur Crewe account books reads:

> 1747. Jan 2nd Pd Mr Bakewell the ironmonger for ye rails at the monument in the Church. £4. 17. 0.

DERBY MUSEUM

In the Local Studies Department of the Derby City Library there is a newspaper cutting headed 'Familiar Derby Inn Sign' and labelled (not dated in the newsprint) December 6th 1938. It reads as follows:

> Those who remember the hanging sign over the doorway of the Bull's Head, Queen Street, which ancient hostelry is about to be demolished for the widening of Full Street will be interested to learn that it is to be added to the exhibits at the Museum in the Wardwick. Not that the painting itself possesses any artistic merit, but the wrought iron frame in which it is enclosed is an excellent example of the work of Robert Bakewell, the clever Derby craftsman who fashioned the Old Silk Mill gates and other greatly prized articles of a similar type. This frame is of some intrinsic value and the Nottingham Brewery Company Ltd. owners of the Bull's Head have kindly presented it to the Museum.

Although it is difficult to compare the flimsy flame tufted frame with other artefacts produced by Bakewell, it is possible to believe that it came from his workshop but was probably made by apprentices.

THE CHAMBERS' MONUMENT, DERBY CATHEDRAL

Near the Consistory is a monument to Thomas Chambers and his wife, who died a short time after the Church, as designed by Gibbs, was completed. It is in marble with two profiles and was sculpted by Roubiliac. The guard rails around the monument are by Robert Bakewell, of course, and lack only judicious gilding to become worthy of their proximity to the Cavendish Screen. Leaving them entirely in black may seem to be appropriate as objects of mourning, but there appears to be a lost opportunity to further enliven the Consistory area of the Cathedral and at the same time draw the attention of visitors to expert craftsmanship.

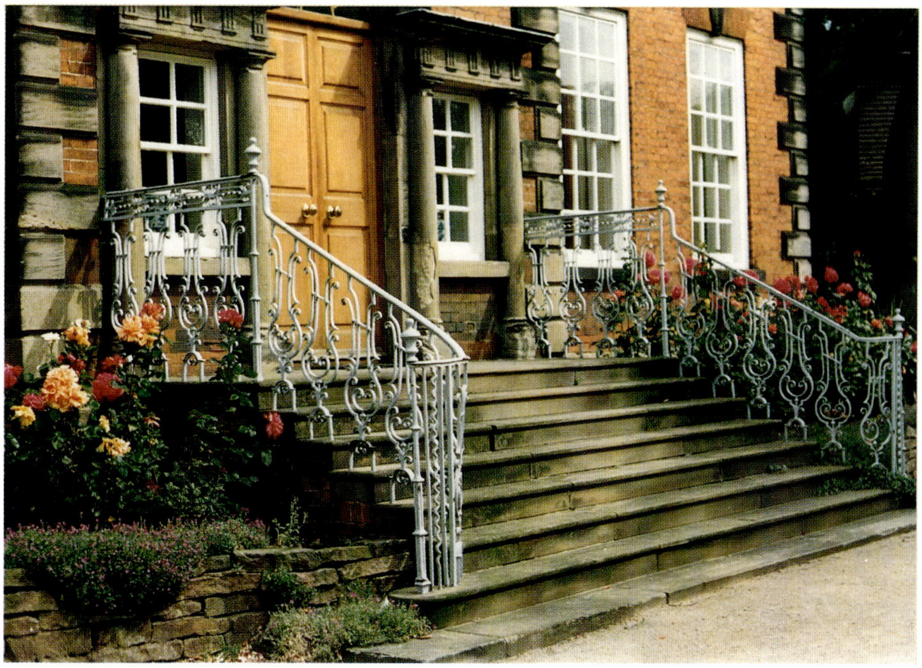

Spondon Derby, The 'Homestead', Perron Ironwork from Darley Hall, Darley Abbey

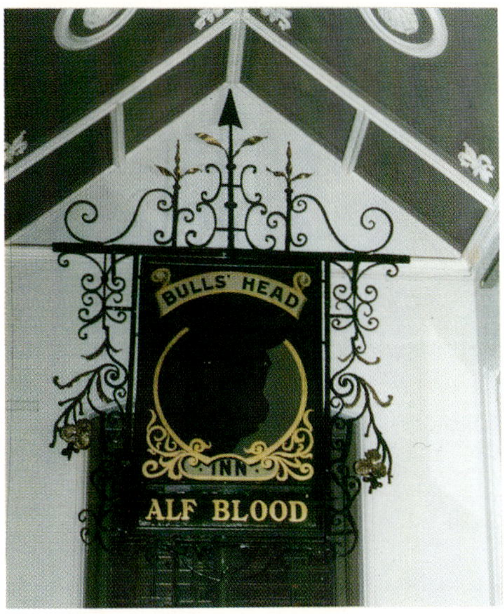

Derby Museum, Inn Sign.

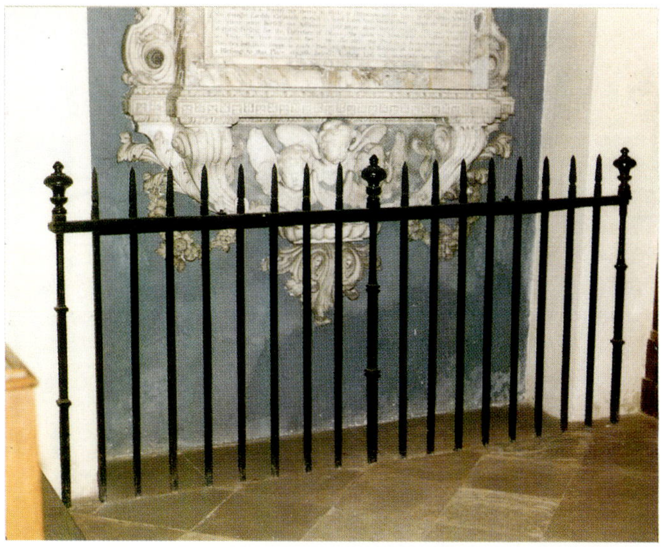

Calke Church, Memorial rails..

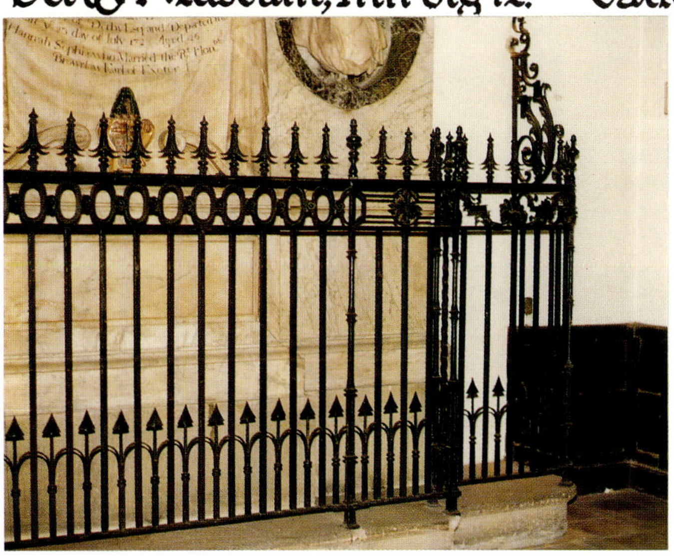

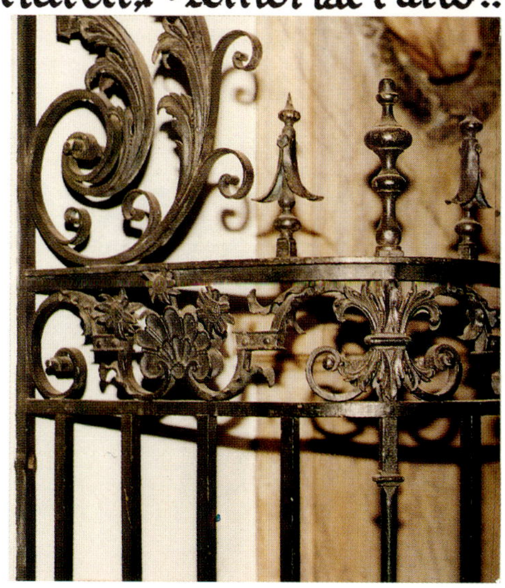

The Cathedral, Derby. The Chambers monument, rails..

LEESWOOD HALL, MOLD, NORTH WALES

The Davies Brothers, blacksmiths of Bersham near Wrexham, practised their craft for many years in the late seventeenth and early eighteenth centuries. They are chiefly remembered for the gates they built for the Castle at Chirk. These startling gates have a mass of repoussé on good ironwork. Some authorities feel that the gates lack planning and style and are too crudely designed. Be that as it may, they show an abandon which in many eyes has its own charm. Because of the popularity of the Chirk Gates it has become the habit to attribute any good gatesmithery in the area to the Davies Brothers. This, in some cases such as the Parish Church, Oswestry; the Parish Church at Wrexham; that at Ruthin; the gates at Emrall Hall and Erddig is quite justified, but only partly so in the case of Leeswood Hall. Rebuilding of the Hall was begun in 1726 for Sir George Wynne who was so fortunate as to have found a lead mine, which at that time, was almost as valuable as one of gold. His temperament being of a prodigal nature, it is said that he paid £1,500 for the Great White Screen shown on Plate 29. (He also had a black screen at the opposite end of the estate).

As to attribution: Ayrton and Silcock in their *Wrought Iron and its Decorative Use, (1929)* refer to "the time when the Davies Brothers and Robert Bakewell worked together". This may appear to be a compromise statement but at the same time we must realise that the colossal size of this screen, very long and tall, was such that one small firm alone could have spent years in completing it. It is highly probable that the two joined forces on the project. There can be little or no doubt that Bakewell had some connection with it, either in a designing or working capacity. Consider the following points; the spear pointed dog bars running the full one hundred and twenty feet length, the inset rectangle in four places on each face of the huge pilasters. The horizontal example of the motif near the base of the pilasters is like the one beneath the terminal bracket at Worcester (Plate 15) and that at the top is a vehicle for the rosette acanthus. This is repeated in a square version at the top of the side frames.

Looking again at the side frame, immediately above the level of the lock bar we see a combination of Bakewell motifs. Starting with the lyre shaped baluster, complete with spur, we come to a version of the finger board construction and end with a second baluster, also complete with spur. The balusters with spurs can be seen in Fig. 3. Adding still more to the Bakewell theme is the diamond strapwork as seen at Aldenham Park (Plate 9). Furthermore, there is the single lock bar arrangement with the bead point fringe along the whole screen except for the two main gates. The finials on top of the huge pilaster caps are reminiscent of those at Henley Hall, Borrowash and Derby Cathedral.

The whole concept of the screen was Palladian and it was probably considered to be ahead of its time.

SAINT OSWALD'S PARISH CHURCH, ASHBOURNE, DERBYSHIRE

No one should ever visit Ashbourne without saving time to visit this Church. Thousands of people do enter the church every year, many to study the three phases of post Norman architecture or just to see the tomb of Penelope Boothby - a child whose effigy was sculpted by Thomas Banks, a forerunner of the Derbyshire genius Sir Francis Chantrey, who sculpted the two children in the Cathedral of Lichfield.

At the East end of the Churchyard are four gates, dwarfed by massive Derbyshire stone pillars each of which is surmounted by a huge obelisk of no light weight resting at each corner on stone skulls. (The word 'macabre' springs readily to mind). In the revision of Pevsner's

Leeswood Hall, Mold,
North Wales.
The White Screen.

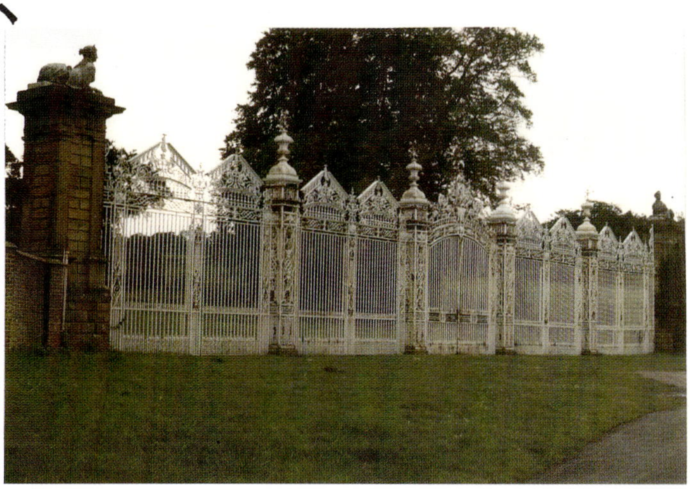

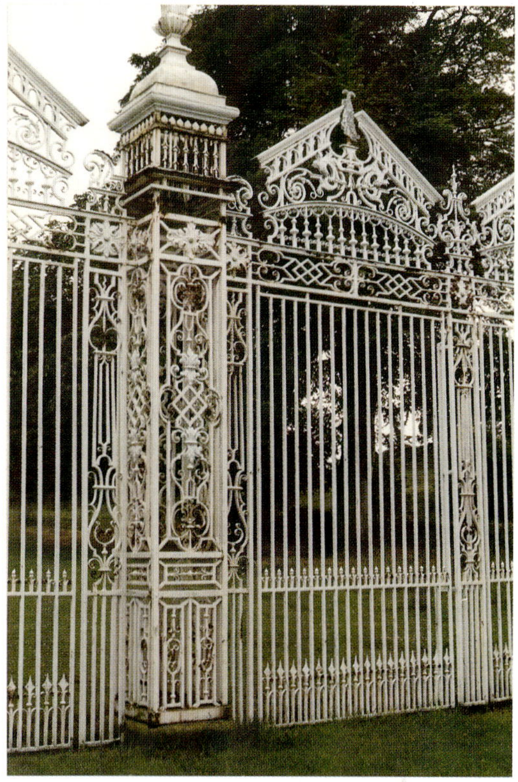

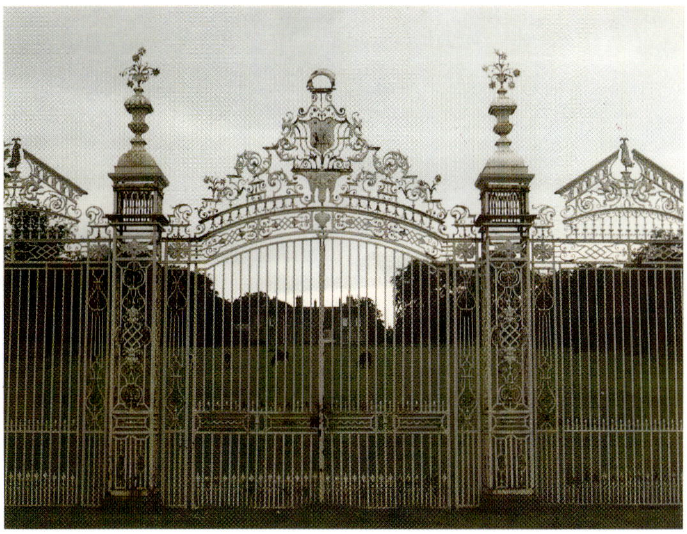

The Parish Church of Saint
Oswald, Ashbourne, Gates...

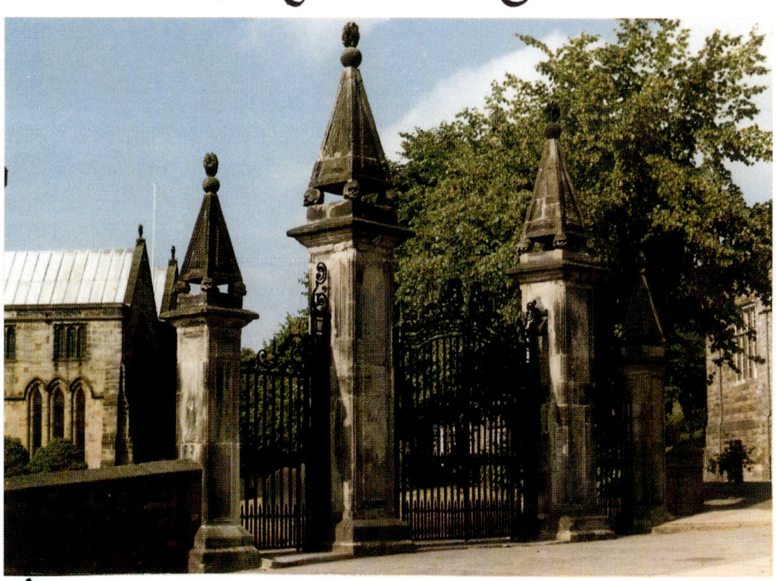

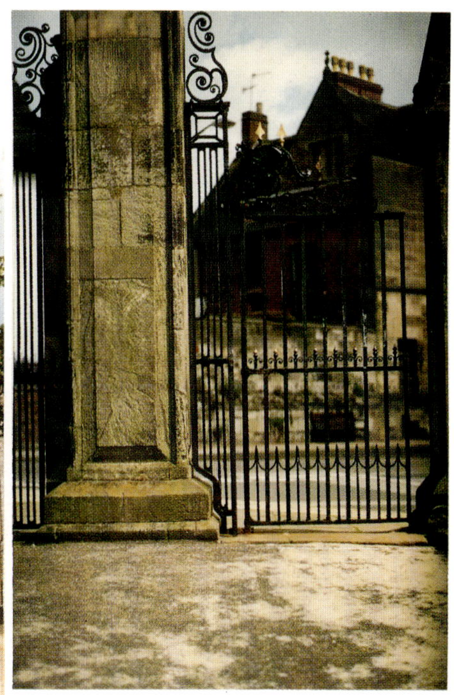

The Buildings of England the editors were advised by the architect and Baroque expert E. J. Saunders, that although there is no documentary proof the gates are "evidently mature work of Bakewell, circa 1730". Justifying Mr. Saunders' view are inset rectangles, close to the top of the side frames, the Bakewell border complete with wavy lines, the arrow pointed dog bars and the bead point fringe on the single lock bar - all regular Bakewell motifs. The terminal brackets, though not strictly identical with others of his, are obviously of the same origin.

CHILLINGTON HALL, WOLVERHAMPTON

The Hall stands in 1,100 acres of Park and woodlands which are so greatly prized as to have been designated a Conservation Area. Tree planting during the period since the second World War has been necessary in order to combat the havoc of Dutch Elm disease, the felling that took place in the war and the fact that so many trees were of such great age. (One had 390 annual rings!). Over two hundred and fifty thousand new trees have been planted since 1950.

Chillington has been the home of the Giffard Family since 1178 -over 800 years. In 1718 Chillington passed to Peter Giffard, who in 1724 commenced the rebuilding of the house. It is believed that the architect was Francis Smith of Warwick. Peter Giffard had his initials along with 'B' for Barbara, his wife, on the heads of the new rainwater pipes together with the date 1724.

A few years later, probably in 1730, was built a huge stone screen having two niches apparently meant for statuary; this was the boundary to a bowling green or alley. (Plate 30). Within the archway was an inner screen of wrought iron. The combination of Bakewell's motifs and the presence of Francis Smith point inevitably to the Smithy at Oake's Yard, Peter Street, Derby, where Bakewell and his men fashioned most of their work prior to transporting it to the required site.

During a recent restoration of the screen at his forge in Leatherhead, Surrey, Richard Quinnell also asserted that he believed the work to be by Bakewell. He arrived at this opinion after noting the lozenge with the wavy line in the stretcher frame and after learning of the history of the work. Plate 30 shows the complete picture and the way in which the stonework dwarfs the ironwork. Centre left reveals the Bakewell border with its wavy line, the tables and the water leaf finials that Bakewell used to achieve the approach to a high centre point. Notice, too, in this picture the swirl or movement in his acanthus. Bottom left shows the arrow pointed dog bars, the bead point fringe on the single bar of the side panels and the top rail of the lock frame together with the Bakewell border within the lock frame all of which support the idea that he was the maker. Bottom right shows that both the narrower side panels have a variation of the finger board construction (Fig. 11) coupled with a suggestion of the lyre form. Top right shows a Bakewell rarity, a mask centred on the stretcher frame. Although the author has heard it treated with a somewhat patronising tolerance, he believes that this is one example of Bakewell's success in a field in which he is supposed to have been particularly limited. Mr. Quinnell informed the author that in the restoration the new masks (one each face of the overthrow) were made of stainless steel. The restoration is logged on the reverse side of each mask. The old masks were cleaned, lacquered, polished and returned to Chillington. Above the mask, in the central cartouche is the cypher in a neat cursive hand.

PLATE THIRTY

Chillington Hall, Codsall Wood, Wolverhampton

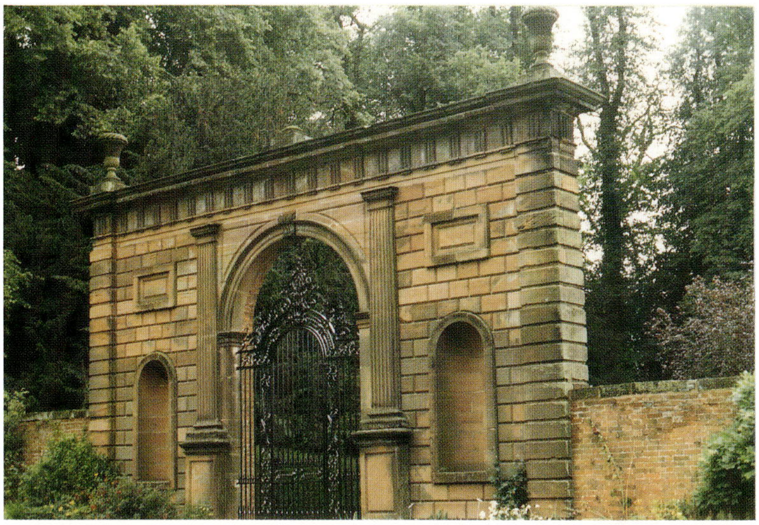

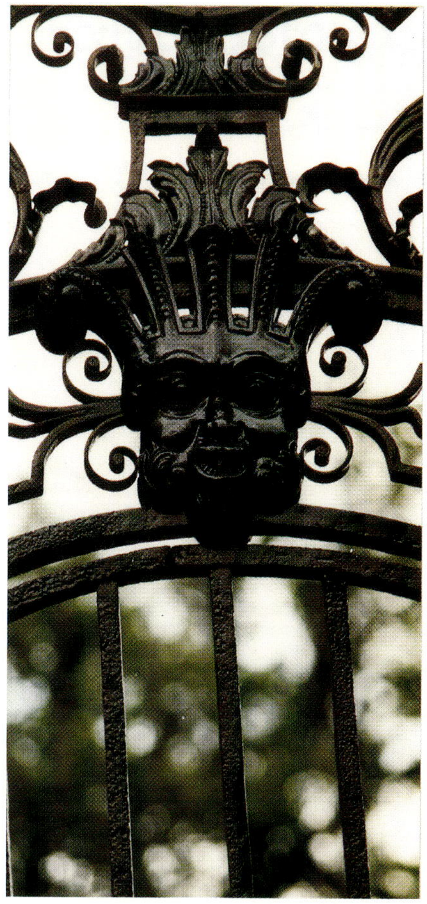

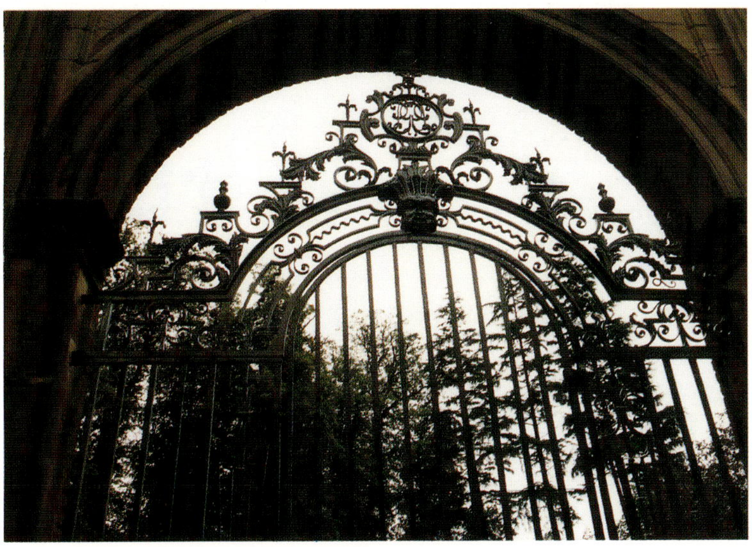

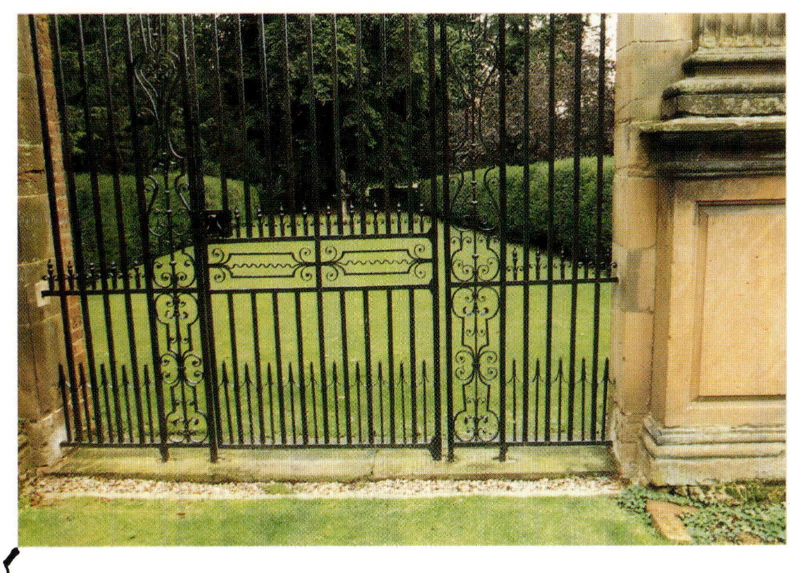

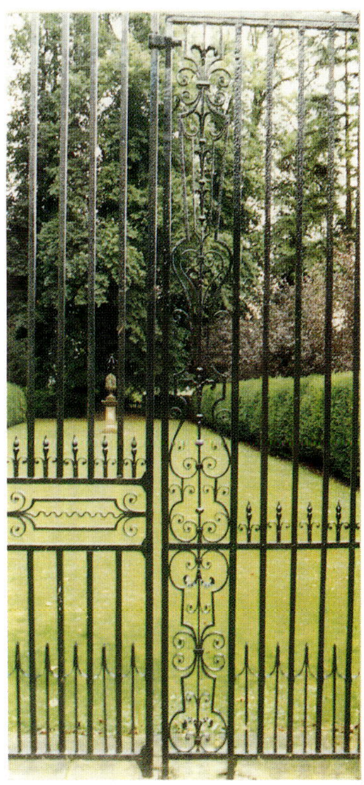

OKEOVER HALL, ASHBOURNE, DERBYSHIRE

The Hall stands on the Staffordshire side of the River Dove, at the entrance to Dove Dale.

In 1729 Okeover Hall came into the possession of Leak Okeover. His Christian name was taken from the maiden name of his mother, Catherine, who was heiress to a comfortable dwelling in Wymeswold, near Loughborough in Leicestershire. Unfortunately Leak was orphaned at the age of twelve years and developed self-willed and expensive habits, which stayed with him far too long. The early eighteenth century was a time when much building and re-fashioning was taking place and Leak Okeover was no exception to those who caught the construction fever. He had very little financial understanding, however, and at one point left for France for a couple of years during which, by dint of the disposal of other properties his trustees managed to pay off his creditors. It is said that he returned somewhat chastened but not cured "either of his gout or his building mania".

Leak had a weakness for wrought ironwork and commissioned Robert Bakewell to carry out a great deal, much of which has been moved around and altered since it was made. Arthur Oswald, writing in *Country Life,* January 1964, reports that Bakewell's bills have been preserved and a selection reads as follows:

An arched gate and all belonging to it	1736
Iron Gates upon the East end of the Terris	1738
Ironwork in the Front	1738
Iron Gates in the Front and also Iron Rales	1739
The new Gates going into the Parke	1740
Iron work in the Return to the Canall	1741
Iron Gates bewist the Grate Peirs	1742

In 1748 Bakewell was responsible for the balustrade to the staircase at Okeover. It is identical with those he made for Staunton Harold (facing Plate 3), and the Maister's House, Kingston-upon-Hull. The only objection to the one at Okeover is that the staircase is too narrow to display the ironwork very well.

Bakewell must have been well pleased with Leak Okeover since he was paid a total of £451 in all - sufficient to keep the wolf from the door for a time. This was not the end of wrought ironwork at the Hall, however, as when Bakewell had done his work Leak turned to Benjamin Yates, who was, after twenty years as his foreman, to become his master's successor. Plate 31 shows Top left, the entrance to the Park at Okeover. The stone pillars remind one of those at Saint Oswald's, Ashbourne, (Plate 29), save for the fact that the obelisks rest upon 'bun-shapes' rather than on skulls. The ironwork is limited to the gates and the overthrow as there are no side panels. Top right shows the gates closed. The dog bars are arrow pointed, the lock frame contains the Bakewell border and is surmounted by the bead point fringe. Centre left, we see more clearly the line of ovals or pseudo ellipses used occasionally by Bakewell (Fig. 14) and the inset rectangle in the centre. Bottom right shows the central cartouche with the family crest and the natural finial with its Oak leaves and Acorns - particularly apt since the family probably took its name from the Oaks in the Park and surrounding areas. Bottom left is a part view of the House gates made by Benjamin Yates. It is included to show how closely he worked to Bakewell's style. Here we see arrow pointed dog bars, the inset rectangle with the wavy line and a variation of the finger board pattern - all of which he must have experienced in various forms over many years.

Okeover Hall, Ilam, Staffordshire. The Park Gates.

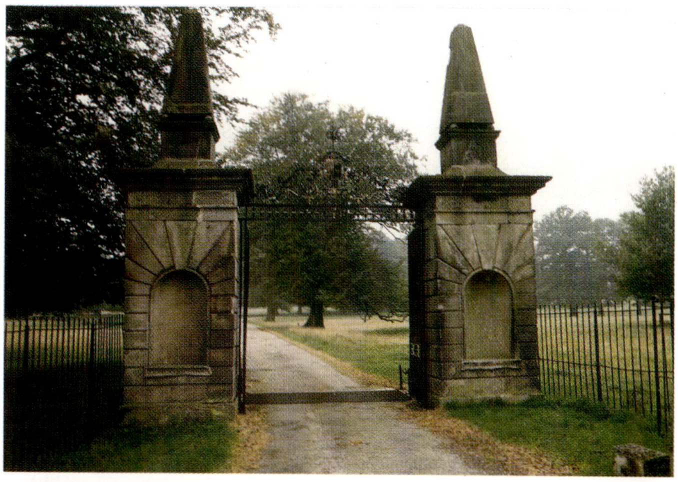

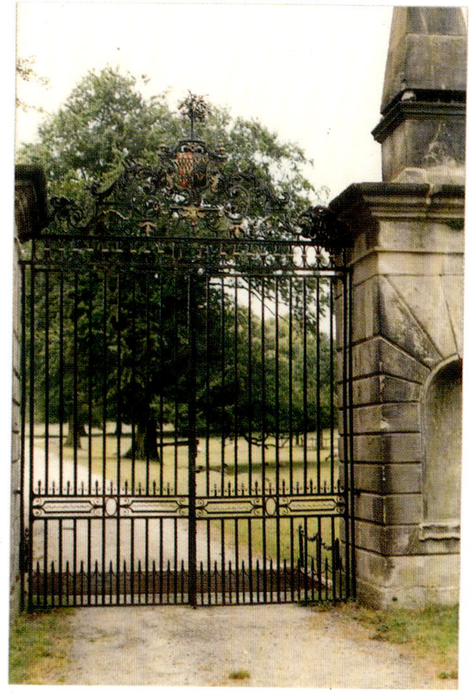

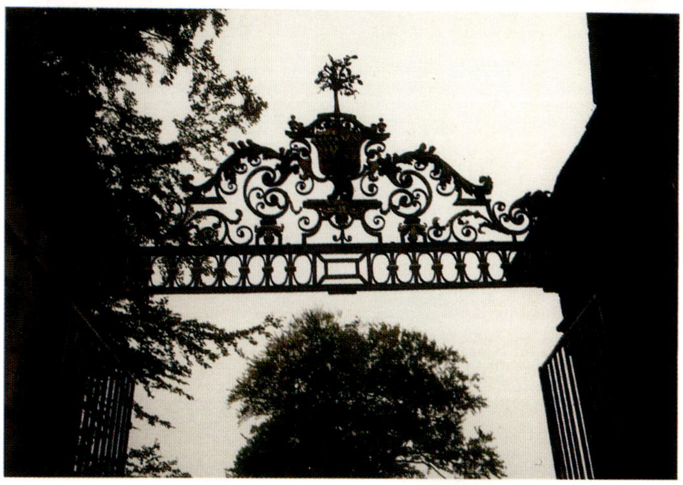

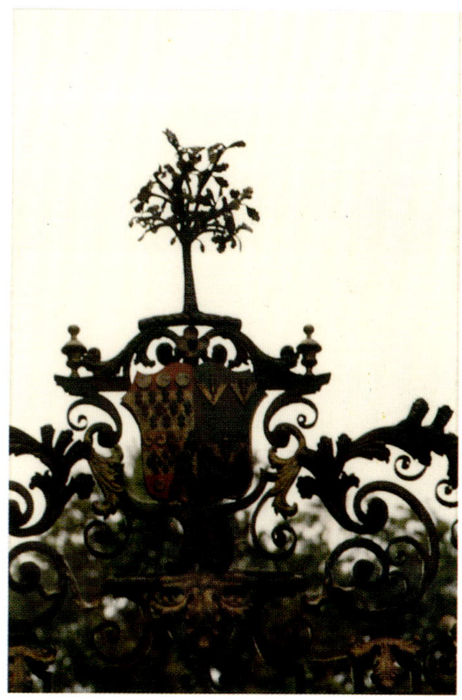

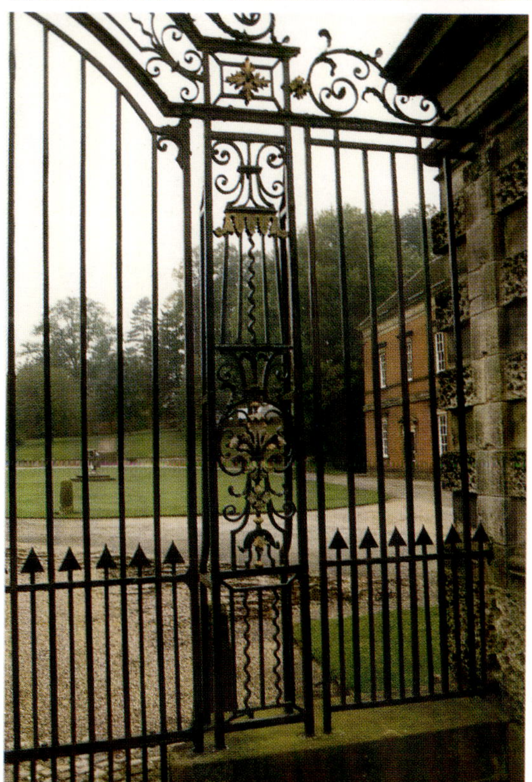

detail of The Hall Gates by Bakewell's former foreman and successor, Benjamin Yates.

THE RADCLIFFE CAMERA, OXFORD

The Camera is a library which has served the University of Oxford for almost two hundred and fifty years. It was designed by James Gibbs, the architect of All Saints' Cathedral in Derby and Saint Martin's-in-the-Fields in London. Gibbs' and Bakewell's work on the Cathedral preceded the Radcliffe Camera by twenty years, which points either to Gibbs's long memory of excellent work or the broadening of Bakewell's reputation.

The building was begun in 1744. The ground floor has sixteen sides and is referred to in official documents as a 'double octagon'. Its overall span is one hundred feet and its greatest height one hundred and forty feet. Thomas Salmon in his publication *The Present State of the Universities*, written in 1744, refers to the building as "the most magnificent Structure in Oxford".

In 1958 S. G. Gillam wrote a full and definitive version of The Building Accounts of the Radcliffe Camera from the accumulated Bills, Minutes and Agreements concerned with the building, and many interesting points are featured. For instance, Bakewell had submitted his tender as follows:

> Proposals for doing the Smiths work of the Radcliff Library at Oxford according to a design of Mr. Gibbs and approved by the Trustees. To fill up one Arch with a circular Head 16½ feet high by 9 feet wide.
> To framing bars of Inch three quarters square.
> To principal bars of Inch and quarter square.
> Circular bars three inches by three quarters.
> Scrolls and bars of Inch eighth.
> The work of each Arch to be about in weight one Tun and a half.
> The work to be done in a Substantial good Workmanlike manner with the best of Swedish Iron, at three pence halfpenny per pound.
> N.B. Three of the Gates to open and no Gottenburgh Iron to be used in the work, all carriage and fixing to be included in this price and to be finished and put up by Michaelmas next
> 1744. Per Robert Bakewell. Recd from Mr. Bakewell 27th Mar 1744.

However, just after Bakewell had submitted his proposals and they had been accepted, war broke out between Britain and France making it necessary to send the ironwork by land rather than by Trent, Coastal waters and the Thames. Also due to risks from French Naval vessels and pirates, the price of iron rose. In his final account, therefore, Bakewell charged fourpence farthing a pound and a special meeting of the Trustees was required before extra payments could be made.

On April 8th 1745 Bakewell received £300 which was his original estimate of the cost and on April 28th 1746 a further £64. 0. 11d., "in full of this Bill and all Demands". The price had risen a little over 20%, but James Gibbs showed his loyalty to Bakewell inasmuch as he reported that the work was "well done" and that "Mr. Bakewell would be a great sufferer if the Trustees tied him down to his first proposal".

The seven grilles are kept in immaculate condition. They are very crowded indeed and are obviously there for protection as well as decoration. In spite of the tight packing they are most handsome. If one looks at the semi-circular head one finds a very different use for spear pointed dog bars. Bakewell used them as a decorative feature pointing towards the centre between the second and third semi-circles.

Bakewell's proposals and the concern shown by James Gibbs that he should be paid in full provide the only documentary evidence to throw some light on the interrelationship between architect and master blacksmith. Communication between Bakewell and his employers, or their builders or architects, was almost entirely verbal. We can only surmise that where he follows his own distinctive style Bakewell was given a free hand with the design, while a noticeable departure from it suggests a need to fulfil the more specific requirements of the person who commissioned the work.

The Plate opposite shows the building, one of the grilles and a more detailed view of the semi-circular head or lunette.

The Radcliffe Camera, Oxford. 1744–46.

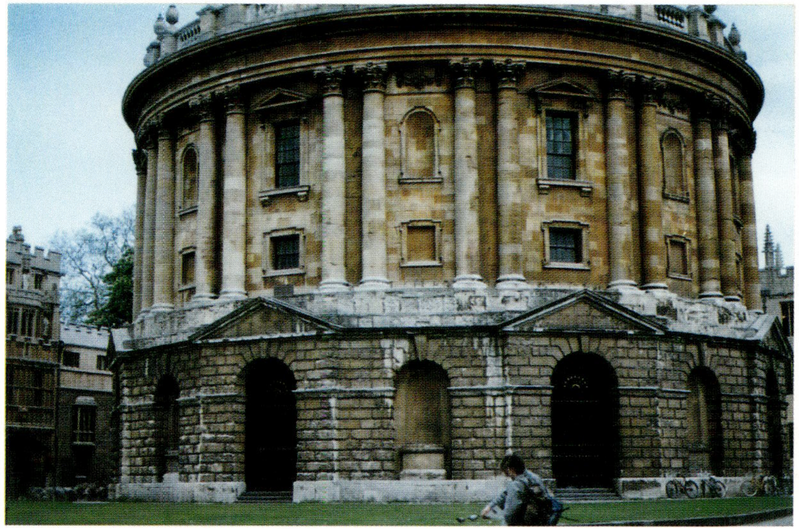

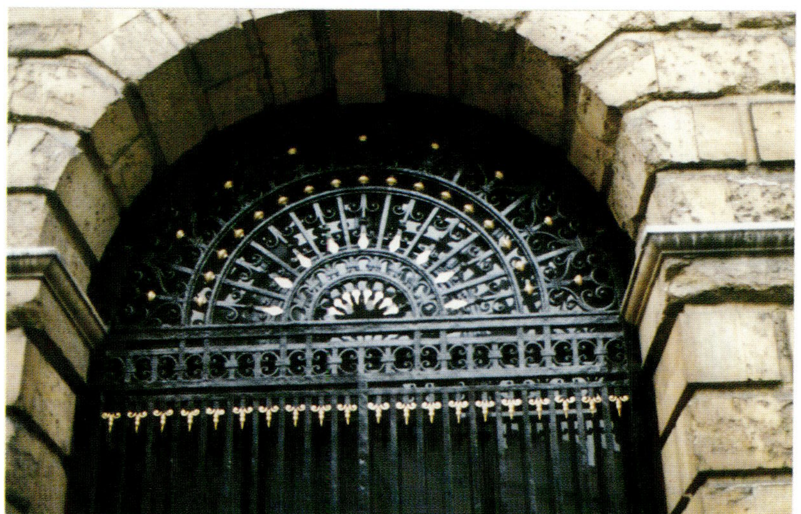

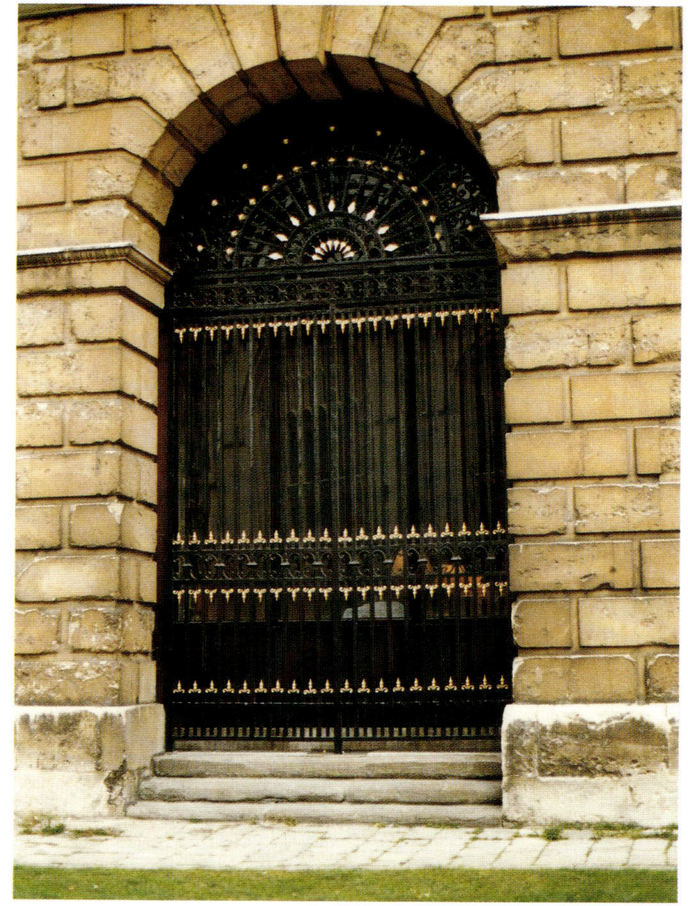

The Seven Iron Grilles in the Outer Arches.

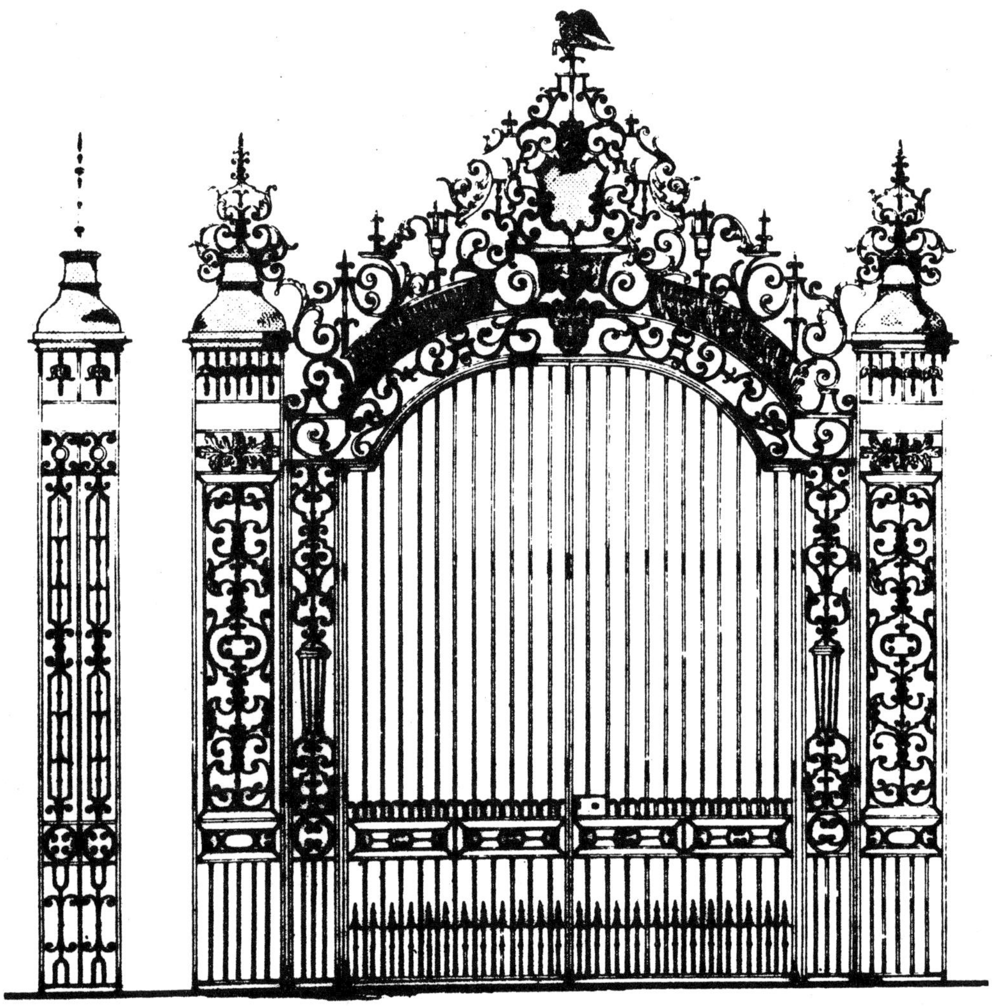

Design of the Combermere Gates praised by Starkie Gardner in the correspondence pages of "Country Life" of November 1919. (See frontispiece and page 30).

CONCLUSION

Although little is known about Robert Bakewell concerning his personality, his physique, his personal relationships or his social life, the lack of information is not unique amongst blacksmiths of his era. He is probably documented more than any of his contemporaries with agreements, bills and so on. Such names as Warren, Gardom, Robinson, Edney, the Paris family, Montigny and the Davies Brothers are all known as craftsmen in iron belonging to the Bakewell times but there is far less known of them even, than of Bakewell.

Marcus Whiffen in his treatise *Georgian Churches outside London* attributes what he refers to as "posthumous anonymity" to the fact that during the English Baroque period the standard achieved by our blacksmiths was so good as to make the work commonplace. It is indicative, therefore, that Bakewell's much better documentation probably springs from the individualism of his general styling and use of motifs. One has to rely to a large extent on this last point since not much of his work was signed with his RB signature unless, of course, more signatures have been lost in restoration. Several things worked in his favour; the times in which he shed the shackles of apprenticeship, thus enabling him to work for himself; the original patronage of Thomas Coke whose position enabled him to refer Bakewell to the right people; and of course his continuing association with Francis Smith the great builder who recognised the worth of Bakewell's work.

Robert Bakewell was a good draughtsman in addition to being a sound craftsman - two features which help each other no matter what material one uses.

As the social life of Bakewell and his family is almost a closed book there is little to be written on the subject. However, we do know the Baptismal dates of his six children as written in the Registers of the Parish Church of Saint Peter in Derby, which are kept in the Derbyshire Record Office at Matlock. They read as follows:

1723.	Bapt. Gratia (Grace)	fil Roberti and Maria Bakewell	28	die July.
1724.	Bapt. Franciscus	fil Roberti and Maria Bakewell	3	die Oct.
1726.	Bapt. Samson	fil Roberti and Maria Bakewell	2	die Feb.
1726.	Sepult. Samson	fil Roberti and Maria Bakewell	22	die July.
1728.	Bapt. Maria	fil Roberti and Maria Bakewell	28	die Marty.
1729.	Bapt. Robertus	fil Roberti and Maria Bakewell	14	die Marty.
1731.	Bapt. Anna	fil Roberti and Maria Bakewell	13	die July.

Samson, the boy named after his paternal grandfather lived only six months. None of Bakewell's children, apparently, had any connection with the blacksmith business.

The last entry in the Saint Peter's Register referring to Bakewell himself reads:

1752. Mr. Robert Bakewell was buried on ye 31st October

Bakewell's popularity may be judged by a letter written by the Reverend Rowland Cotton of Etwall, (see Plate 23 and text and dated December 6th 1749, to his friend the Dean of Exeter in which he says:

> "I found on my return a Draught of the West Door of Tutbury Church; it seems to be accurately don but as the artist has not observed my order in discribing the Ornaments only of one half, he has made a more operable affair of it than was necessary. This country does not produce many such artists: It was some months before I could find one that durst undertake it! I apply'd first to Bakewell, famous all over England for Ironwork and who Drawes to perfection but he declined it, I believe, upon the account of his Eyes wch begin to fail him".

It was possibly at this time, (1749 -50) that Benjamin Yates took over the management of Bakewell's business. He had been his foreman for eighteen years and for some years before he was his apprentice, and must have been thoroughly accustomed to his designing and his methods of production.

On Bakewell's demise Yates moved from the smithy in Oake's Yard, Derby which his master had occupied for forty years, to a new shop in the Wardwick, Derby, near the Church of Saint Werburgh. He announced that he had taken over the business in *The Derby Courier*, December 1752.

ADVERTISEMENTS.

BENJAMIN YATES, *Irongate-ſmith,*

(Who ſerved an Apprenticeſhip, and was after, for upwards of twenty Years, FOREMAN *to the late ingenious Mr.* ROBERT BAKEWELL, *of the ſame Buſineſs in* DERBY, *Deceaſed,)*

HEREBY GIVES NOTICE,

THAT he intends to carry on the ſaid Buſineſs, at his New built SHOP in the *Wardwick*, near St. *Werburgh's* Church in DERBY, where any Gentlemen may have IRON GATES, PALLISADOES, &c. for Gentlemens Seats, and other Publick Buildings; with all the ſeveral Branches thereof, perform'd in the compleateſt Manner; as was done during the Life-time of his Maſter:

By their humble Servant,

B. YATES.

In turn he was succeeded by a relative, William Yates, in 1776.

This, then has been the bare record of a man who gave something very English to the Baroque movement in this country, and became the leading native-born smith of his time. Much has been made of the fact that his grave is unmarked. The author believes that it lies beneath Saint Peter's Churchyard in Derby. The only memorial, as such, to be placed anywhere is on the Stone House at Melbourne - probably, apart from his grave, the most suitable place for it.

It was said of Sir Christopher Wren that those who wished to see his memorial should simply look around them. Surely we can think the same of a man who produced the Cathedral Screen, the Worcester Guildhall Gates and Rails, the Garden Arbour at Melbourne and all the other charming and dignified suites of Gates and Rails that have been illustrated in the foregoing pages.

The Anatomy of Carriage Gates.

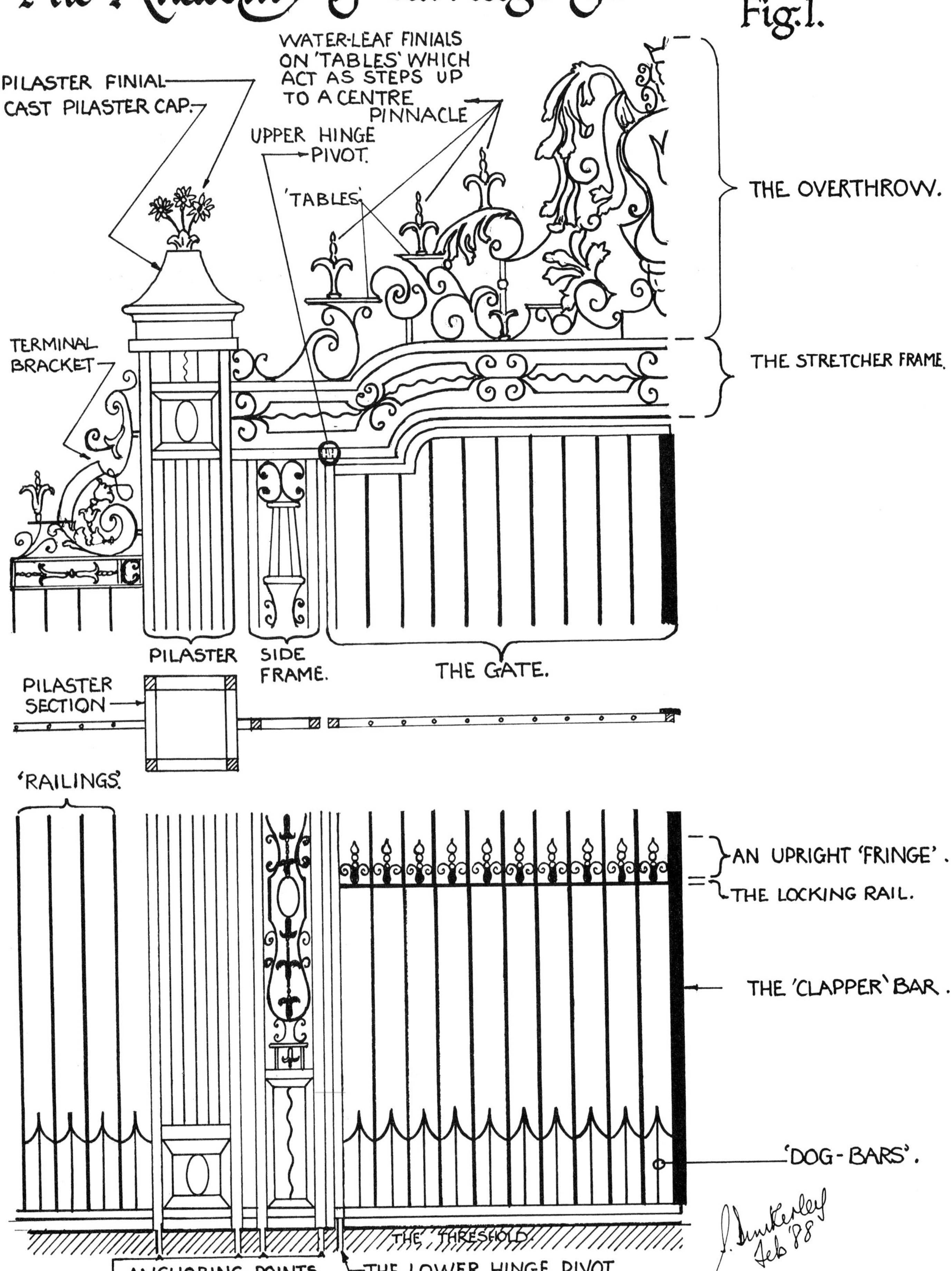

Pillars, Piers and Pilasters..

Fig. 2.

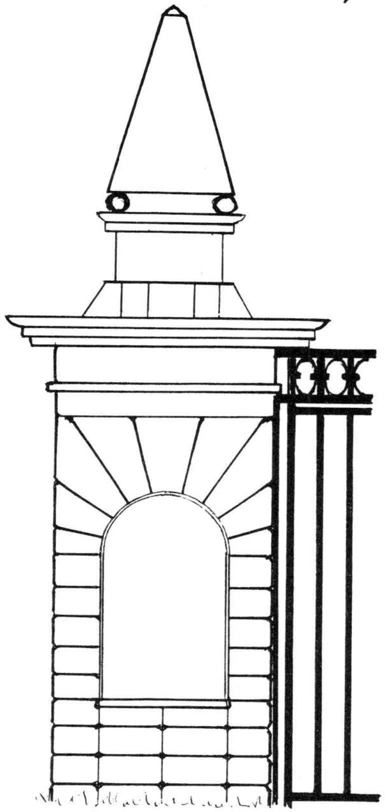

HEAVY STONE PILLAR. GATE HINGED DIRECTLY NEXT TO THE PILLAR.

AS AT OKEOVER.

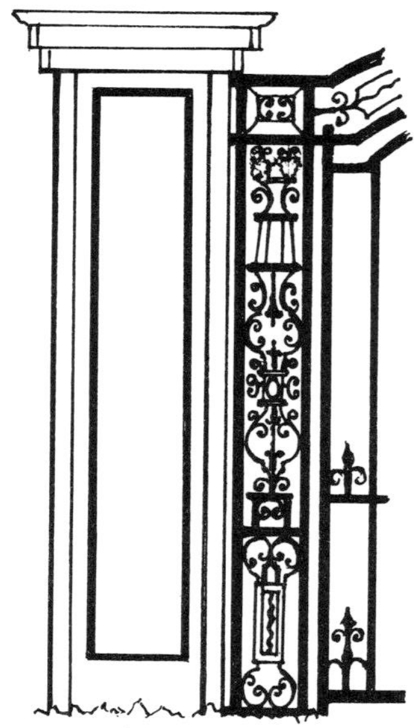

STONE PILLAR. 'SIDE' FRAME, SHOWING MANY BAKEWELL MOTIFS SEPARATES GATE FROM PILLAR.

AS AT CHURCH LEIGH.

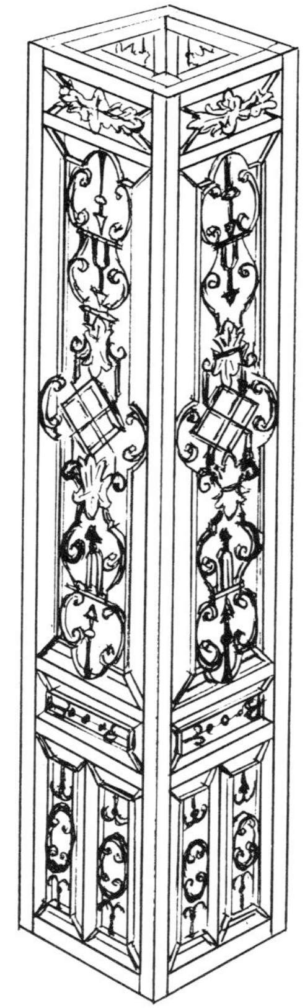

WROUGHT IRON PIER. EACH OF THE FOUR ELEVATIONS IS COMPLETELY DECORATED.

AS AT LEESWOOD.

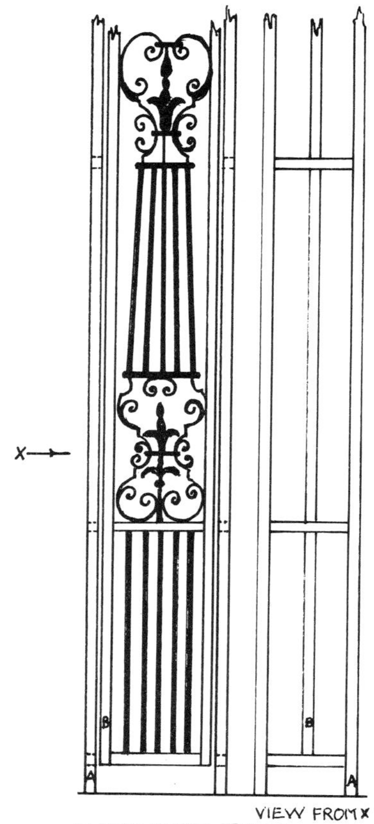

WROUGHT IRON PILASTER. THE DECORATION IS CONFINED TO A SEPARATE PANEL. (B)

AS AT PENSHURST.

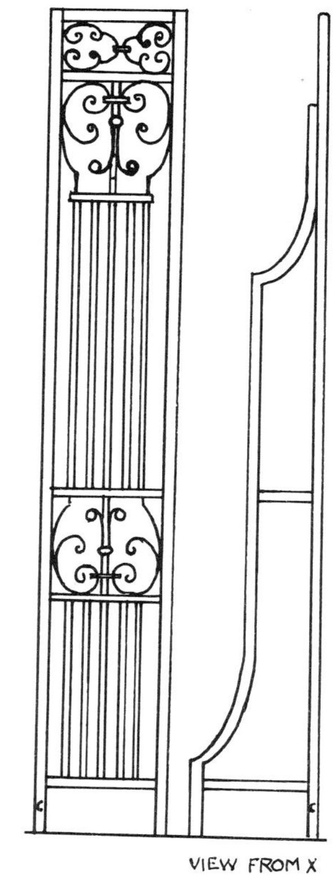

PROBABLY THE CHEAPEST TYPE OF BAKEWELL PILASTER.

AS AT ETWALL.

Bakewell balustrades: Fig. 3.

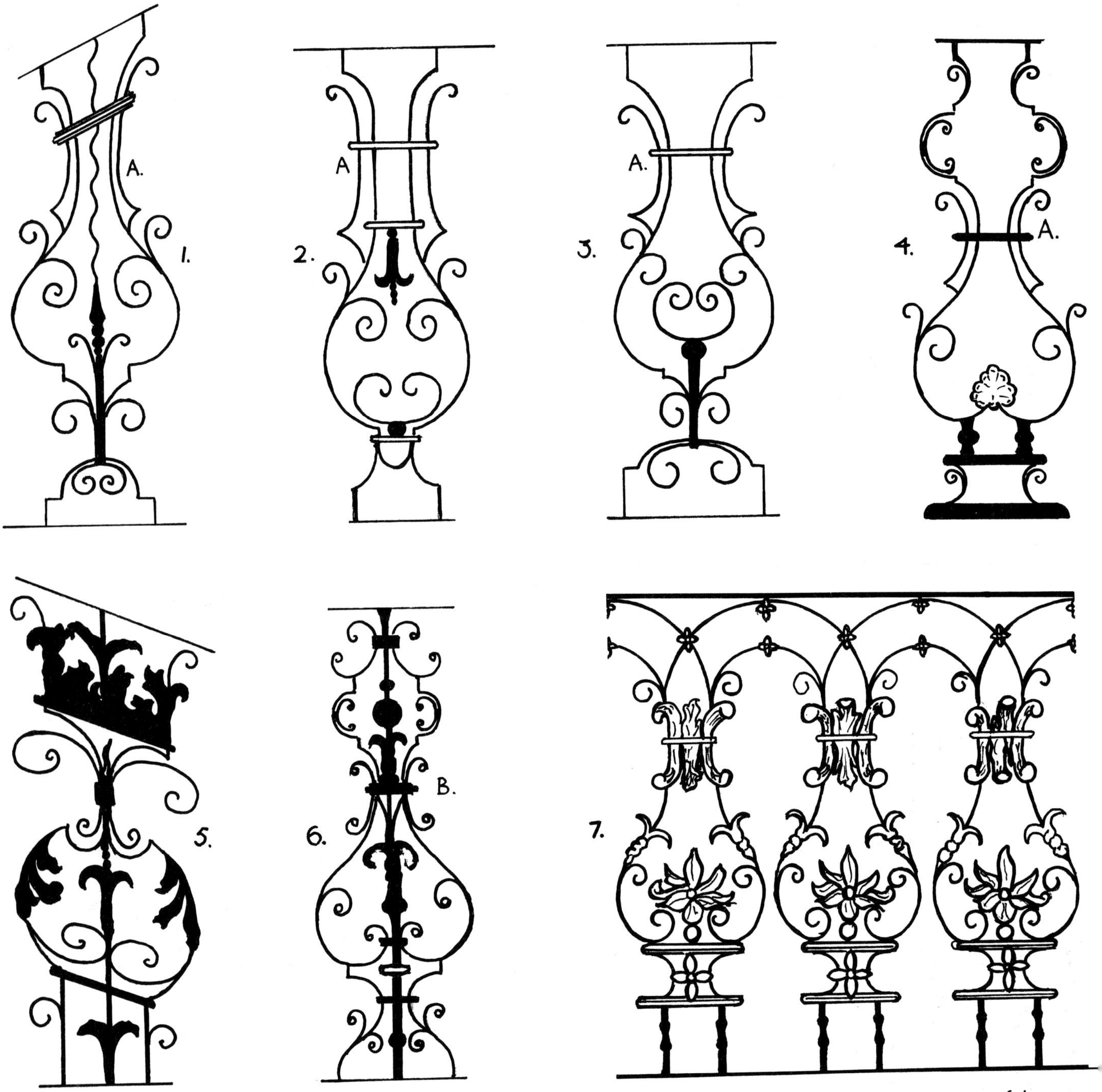

THE 'SPUR' AT 'A' IN FIGS 1-4 IS A BAKEWELL 'SIGNATURE', REPLACED BY A 'C' SCROLL IN 6 (B).

1. TOWNELEY HALL ART GALLERY, BURNLEY. THE STAIRCASE.
2. PERRON IRONWORK, ST. MARYS GATE HOUSE, DERBY. (DEMOLISHED).
3. PERRON IRONWORK, THE HOMESTEAD, SPONDON. FORMERLEY AT DARLEY ABBEY.
4. THE GARDEN BALUSTRADE, THE HALL, MELBOURNE, DERBYSHIRE.
5. CHOLMONDELEY CASTLE NEAR MALPAS, CHESHIRE. THE STAIRCASE. (IN FRENCH TASTE.)
6. ALL SAINTS' CATHEDRAL, DERBY. THE COMMUNION RAILS. BLUE AND GOLD.
7. 'LINKED BALUSTERS' SIMILAR TO THOSE AT STAUNTON HAROLD, OKEOVER AND THE MAISTER'S HOUSE IN HULL.

J. Dunkerley
Jan '88

Anti-dog devices. 'Dog-bars.' Fig. 4.

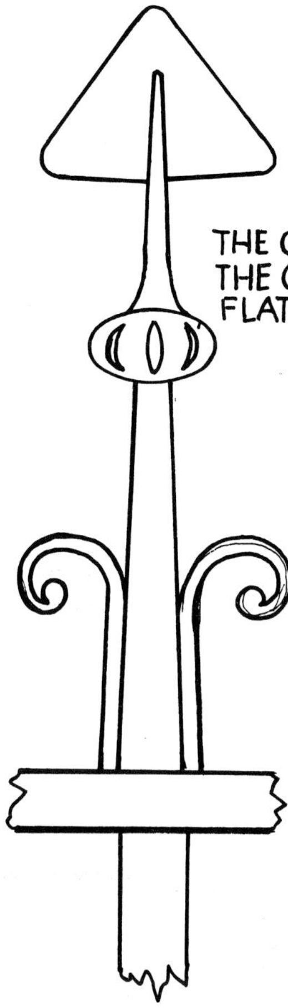

THE CATHEDRAL, DERBY. THE OSBORN GATES. FLAT ARROWHEADS.

←(LEFT) SPEARPOINTS AS AT PRESTON-ON-WEALD.

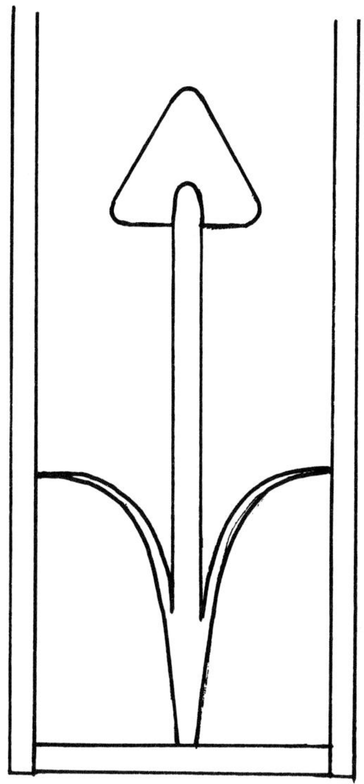

THE CATHEDRAL, DERBY. THE CHAMBERS MEMORIAL, FLAT ARROWHEADS.

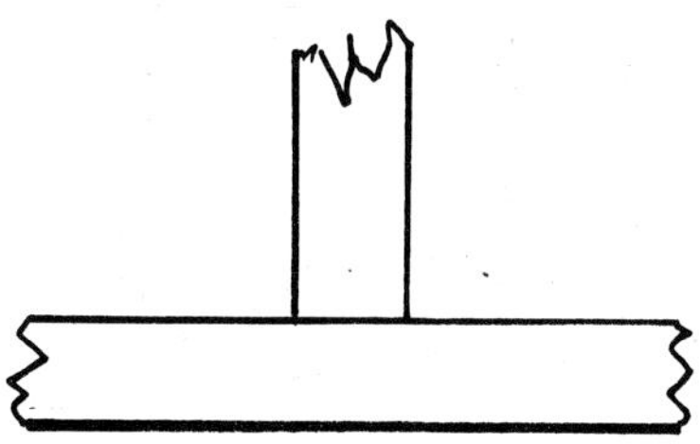

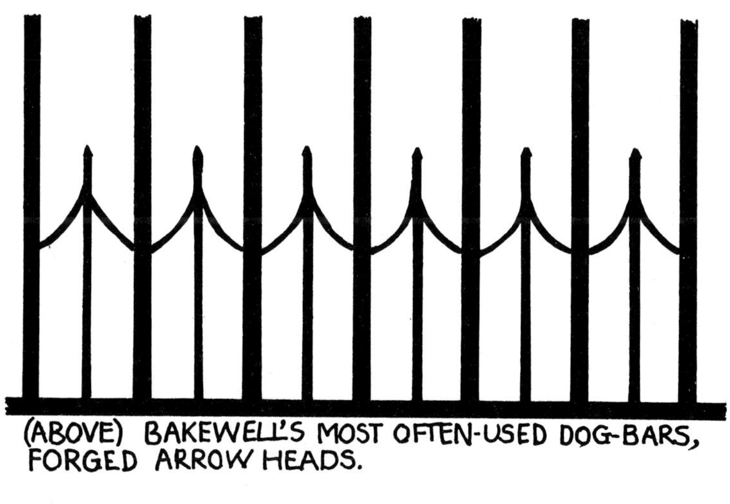

(ABOVE) BAKEWELL'S MOST OFTEN-USED DOG-BARS, FORGED ARROW HEADS.

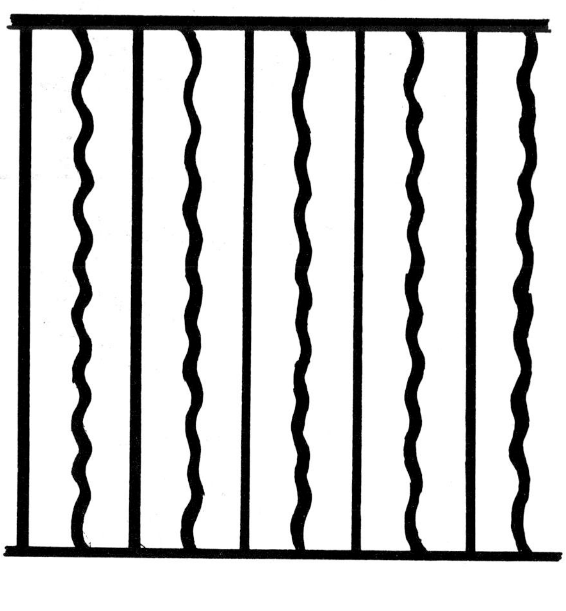

(LEFT) FOREMARK AVENUE GATES AND RAILS. S. Dunkerley. Jan '88

'thunder and lightning bars.'

Crowning the Pilasters Fig. 5.

AT THE ALMSHOUSES, ETWALL, DERBY.

AT PENSHURST PLACE, KENT, EX WINGERWORTH, DERBYS.

AT BIRD STREET, LICHFIELD.

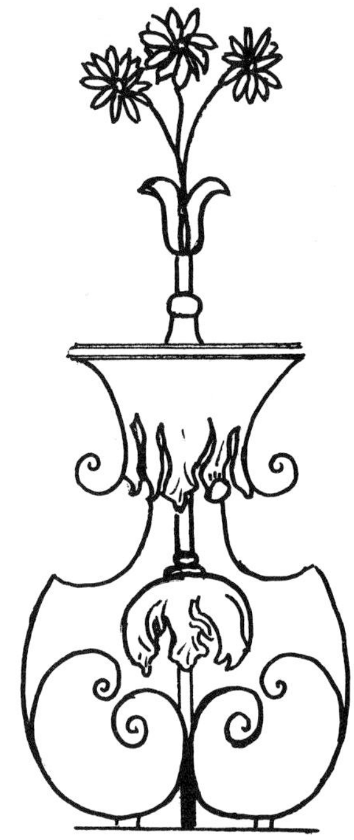

AT HENLEY HALL, LUDLOW.

and 'standards'.

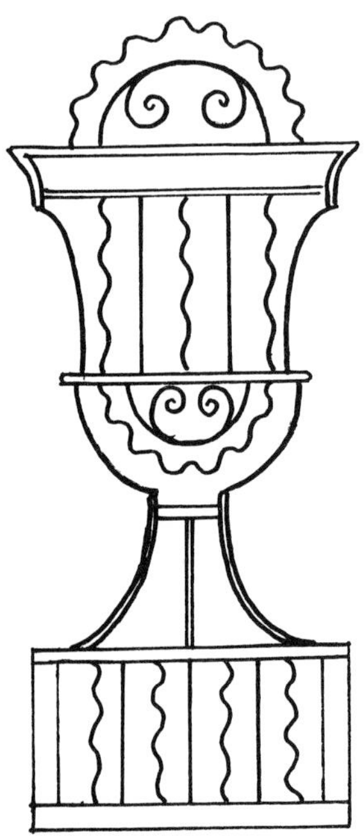

AT HIGH GREEN, CANNOCK.

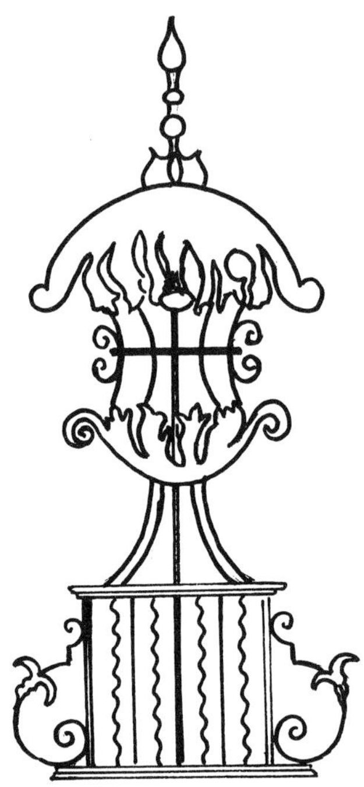

AT ALL SAINTS CATHEDRAL, DERBY & NEWNHAM PADDOCKS, LEICESTERSHIRE.

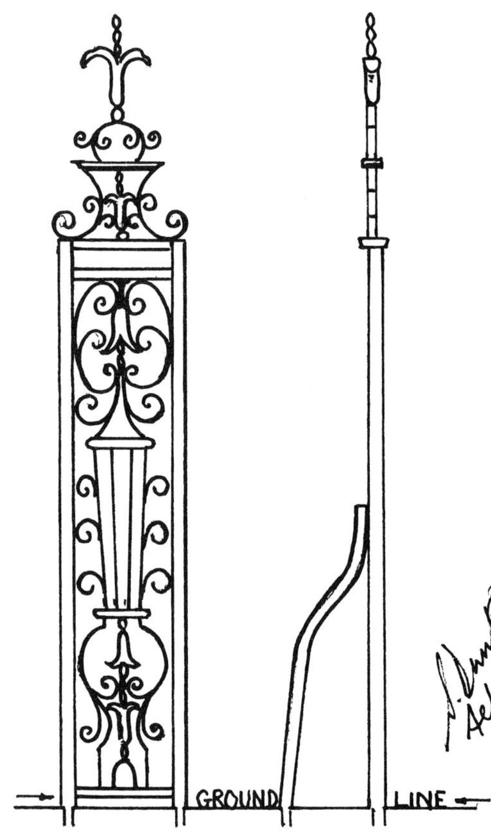

STANDARDS, TOO, ARE 'CROWNED' AS AT WORCESTER GUILDHALL.

Bakewell motifs: the "inset rectangle", variations. Fig: 6.

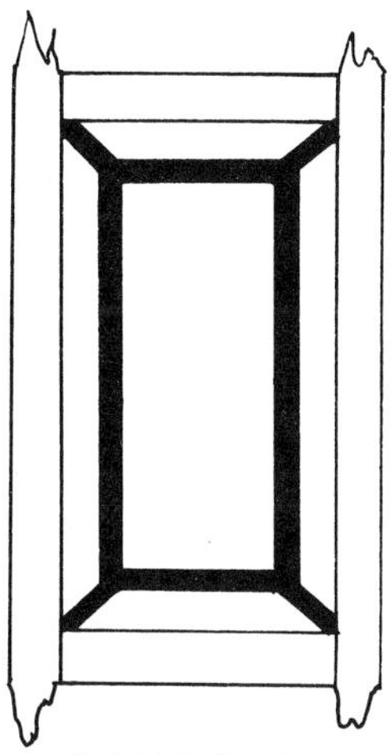

ASHBOURNE
CHESSHYRE GTE.
MELBOURNE.
OKEOVER
WORCESTER.

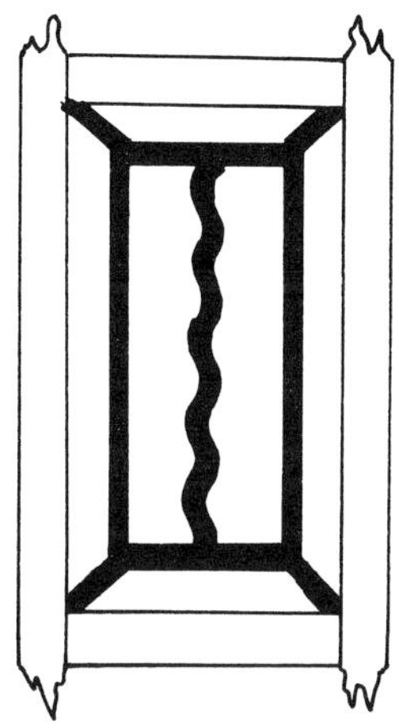

OSBORNE GATES.
CATHEDRAL SCREEN.
CANNOCK.
CHURCH LEIGH.

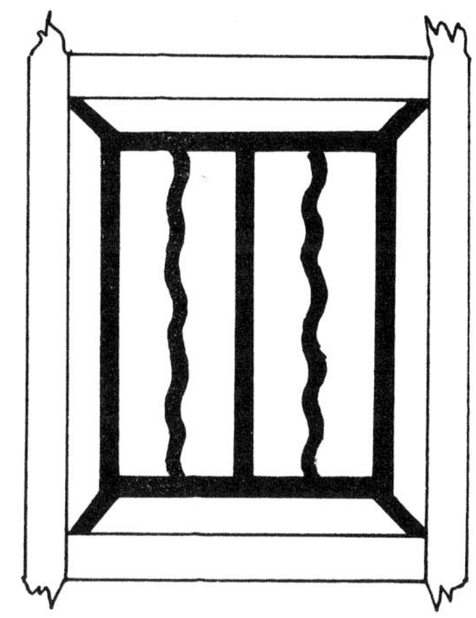

CHESSHYRE GATE.

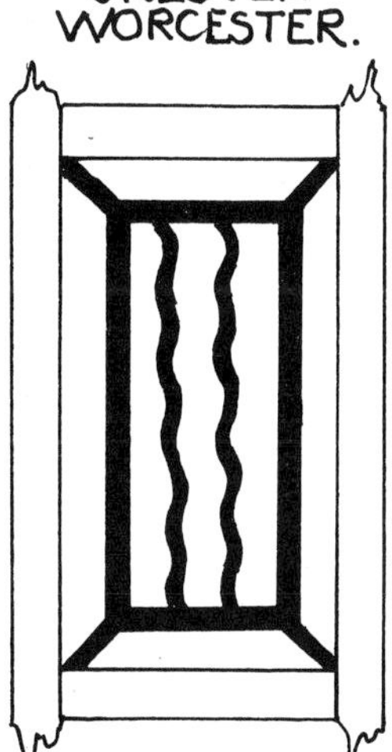

FOREMARK, THE AVENUE GATES.

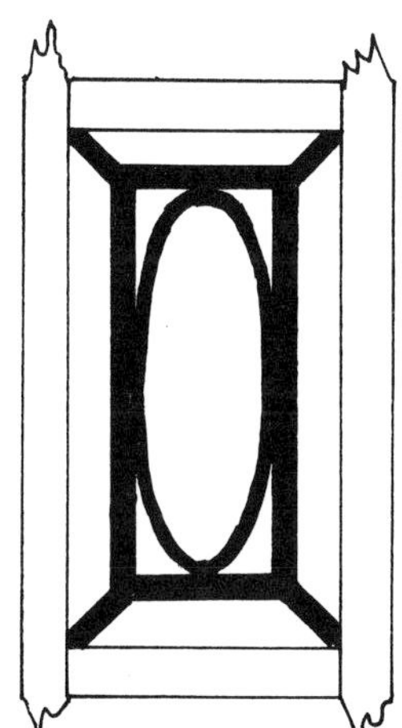

CATHEDRAL SCREEN, DERBY.

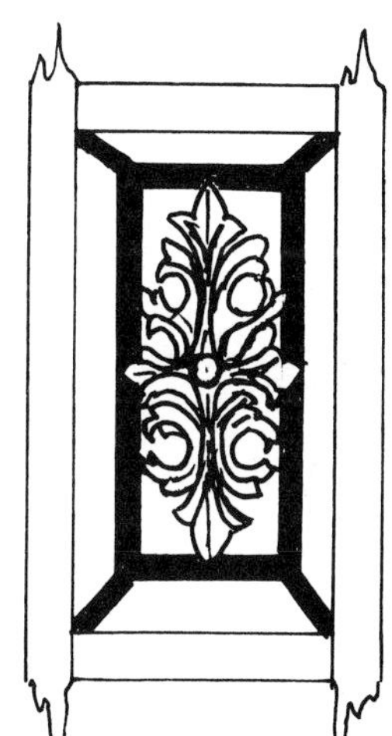

DERBY CATHEDRAL.
LEESWOOD.
OKEOVER.
WORCESTER.

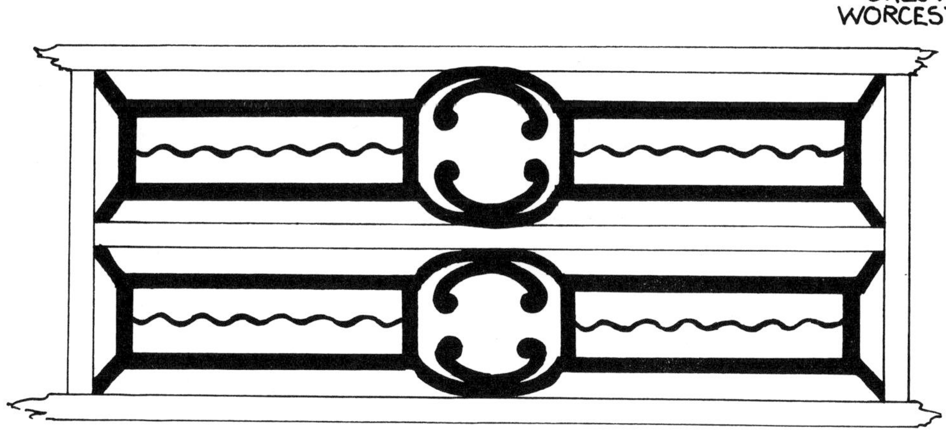

THE CATHEDRAL SCREEN, DERBY, BUT PLACED VERTICALLY.

Bakewell motifs:

Fig. 7.

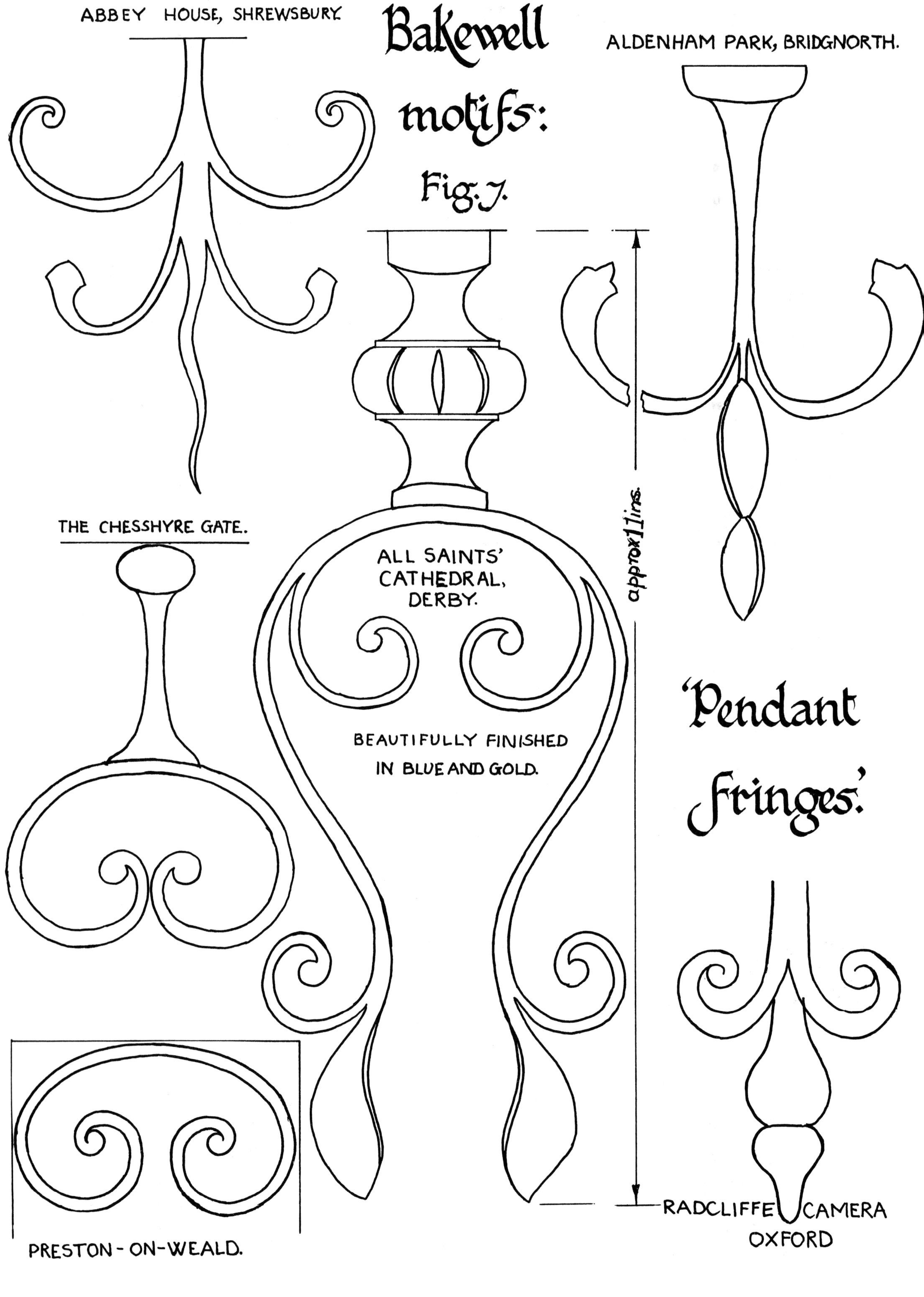

'Pendant Fringes.'

'Reaching for the Sky.'

Fig: 8.

The 'upward sweep'.

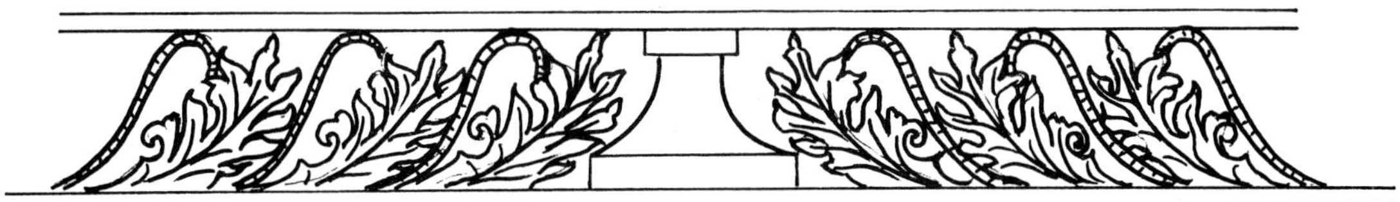

ABOVE: TIJOU'S IDEA TO CREATE A SENSE OF UPWARD SWEEP TO A PINNACLE.
BELOW: BAKEWELL USED BEADED 'POINTS' TO ACHIEVE THE SAME EFFECT ON THE SILK MILL, ASTON AND ALDENHAM PARK GATES. HIS LINES OF OVALS AT ALDENHAM WERE USED TO ACCENTUATE THE VERTICAL.

Crestings.

ON SINGLE GATES THE HIGH POINT IS CENTRAL

AS AT LUDLOW

ON DOUBLE GATES THE HIGH POINTS ARE ON THE CLOSURE SIDE CREATING THE ILLUSION OF AN OVERTHROW

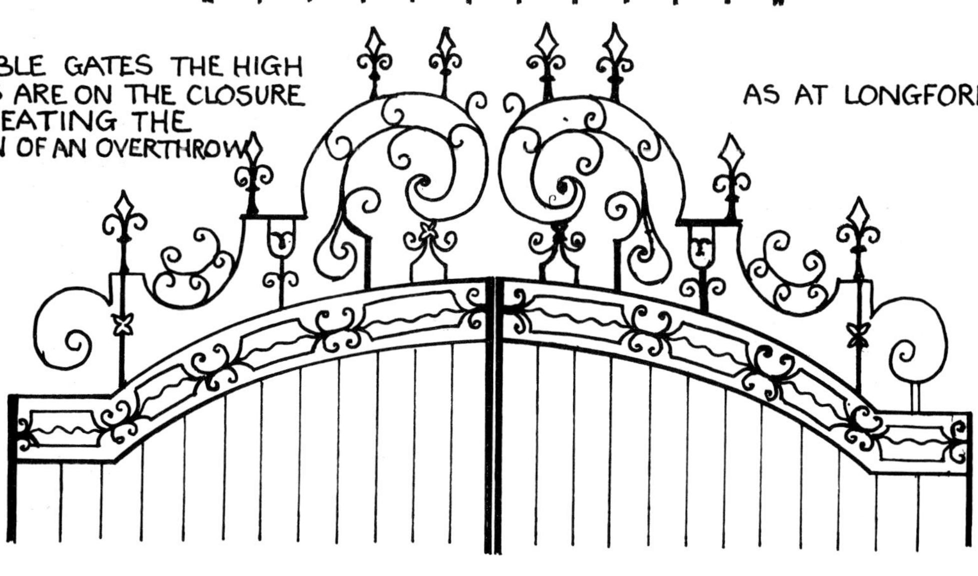

AS AT LONGFORD.

'Reaching for the Sky.' Fig. 9.

THE LOWEST FRAME—PERFECTLY FLAT—AS AT ETWALL, ALDENHAM & DERBY CATHEDRAL.

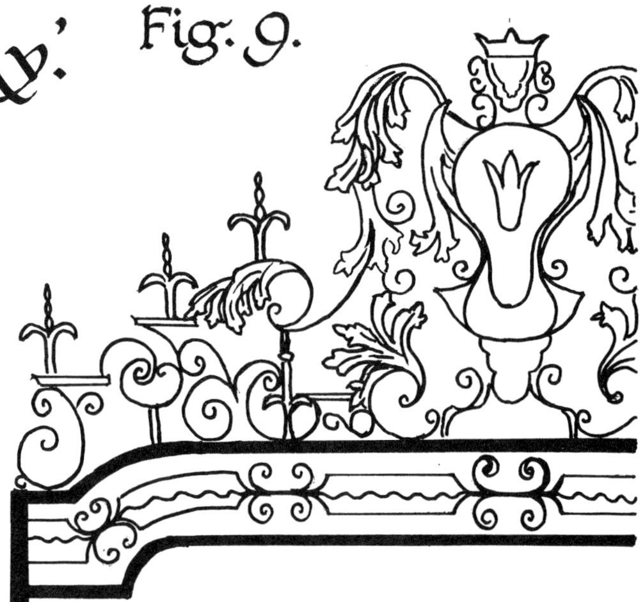

THE FIRST STEP UP—THE RAISED FRAME—AS AT PENSHURST PLACE & LLANRHYDD.

(ABOVE) THE SECOND PROGRESSION—THE DEPRESSED ARCH, SEEN AT CHURCH LEIGH AND WORCESTER.

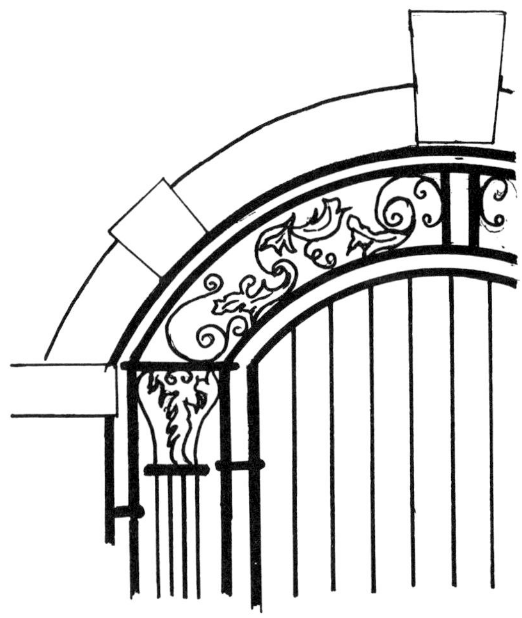

(ABOVE) THE THIRD PROGRESSION—THE STEEPLY-ARCHED FRAME AS AT TISSINGTON, CHILLINGTON AND CANNOCK.

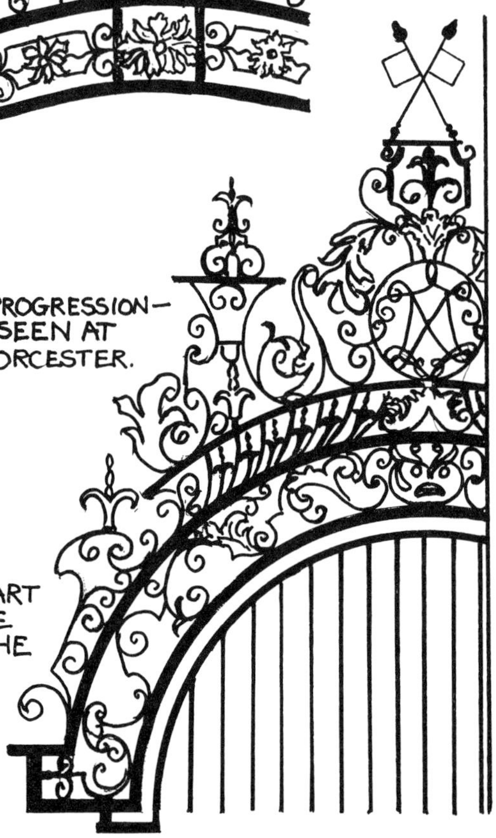

THE HIGHEST STEP APART FROM THE GOTHIC—THE SEMI-CIRCLE AS AT THE DERBY SILK MILL.

stretcher or overthrow-base frames.

THE MAKING OF A 'CYPHER'

'J' FOR 'JOHN', 'H' FOR 'HEATH', THE MONOGRAM, 'J' REVERSED, 'H' REVERSED, THE MONOGRAM REVERSED, THE COMPLETE 'CYPHER'.

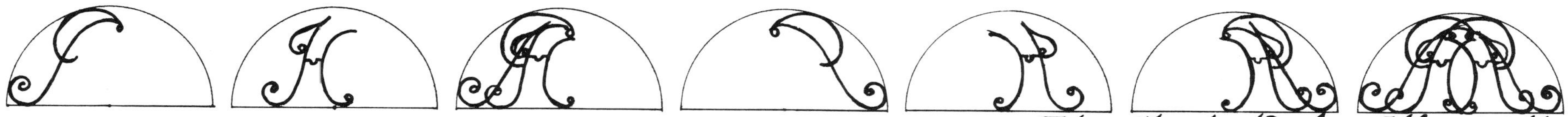

John Heath, Derby Silk Mill.
The 'Darwin' Fanlight.

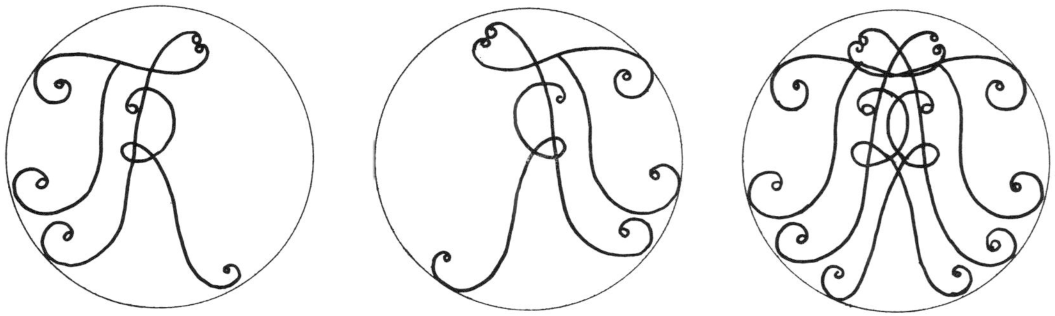

Thomas Rivett, Derby.

Fig: 10.

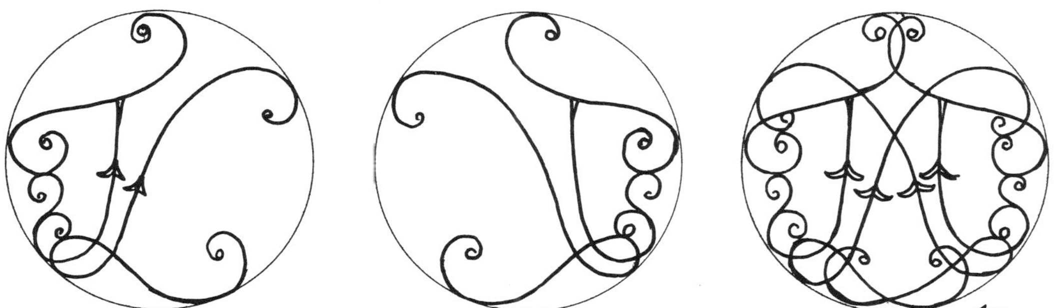

Thomas Lombe, Silk Mill Gates.

Examples of Bakewell Cyphers.

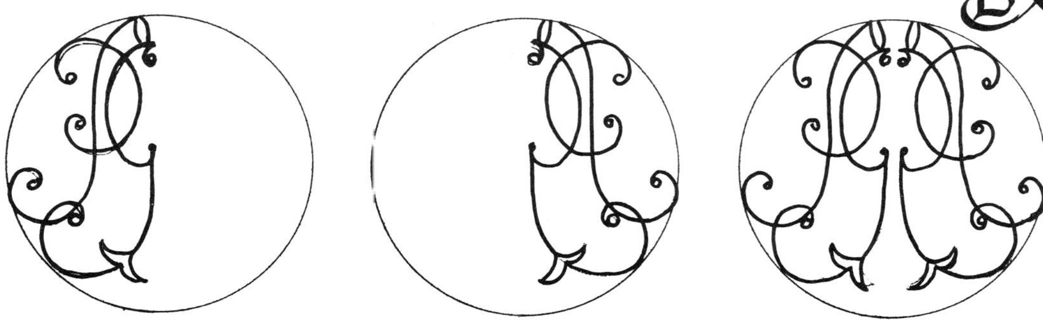

Peter Giffard, Chillington Hall

J. Dunkerley 1st Mar. 1988

Bakewell motifs: Fig. 11.

Variations on a theme.

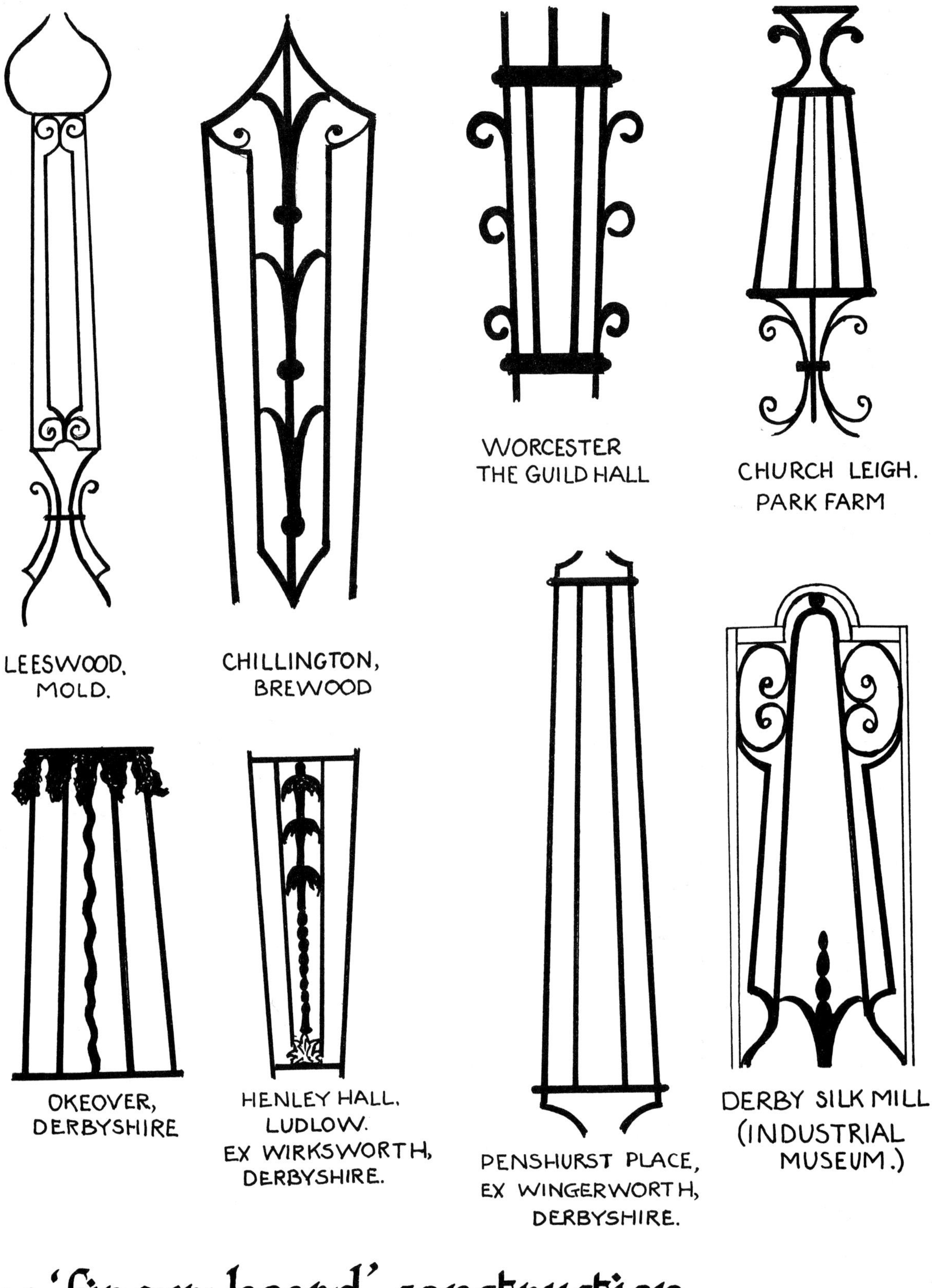

the 'finger-board' construction.

Bakewell motifs: linear patterns Fig: 12

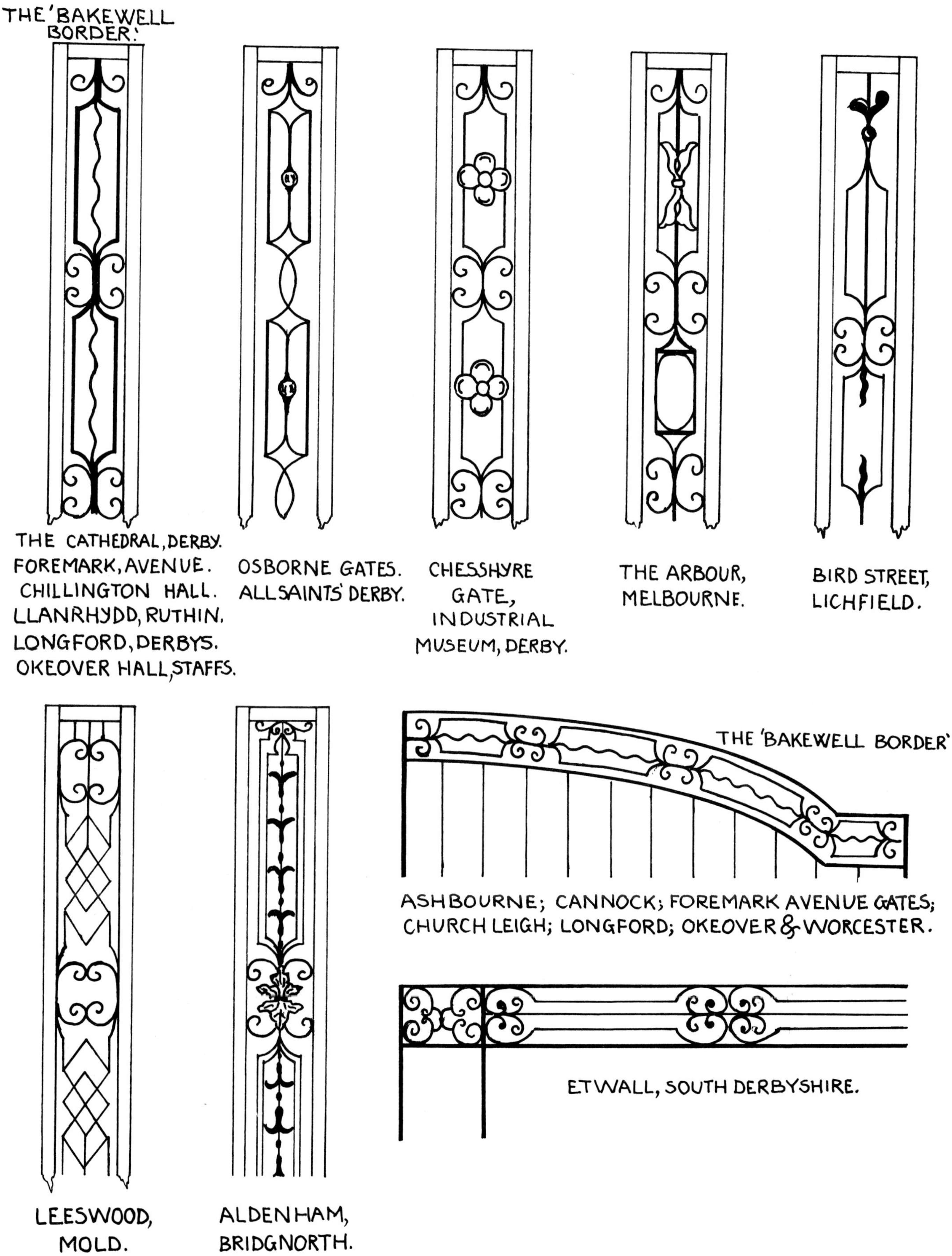

Bakewell motifs: 'Upright Fringes'

Fig. 13.

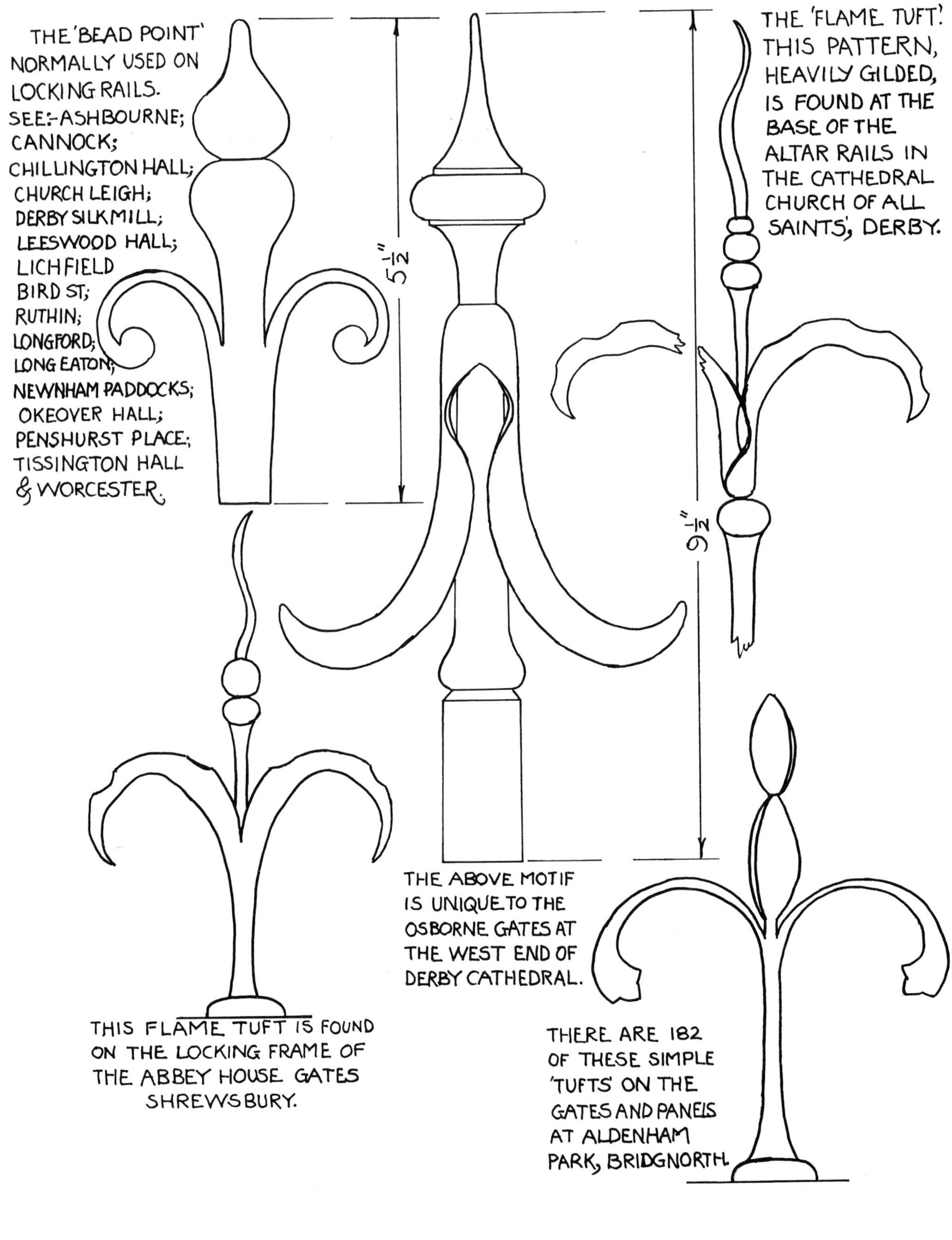

Bakewell motifs: the ellipse. Fig.14.

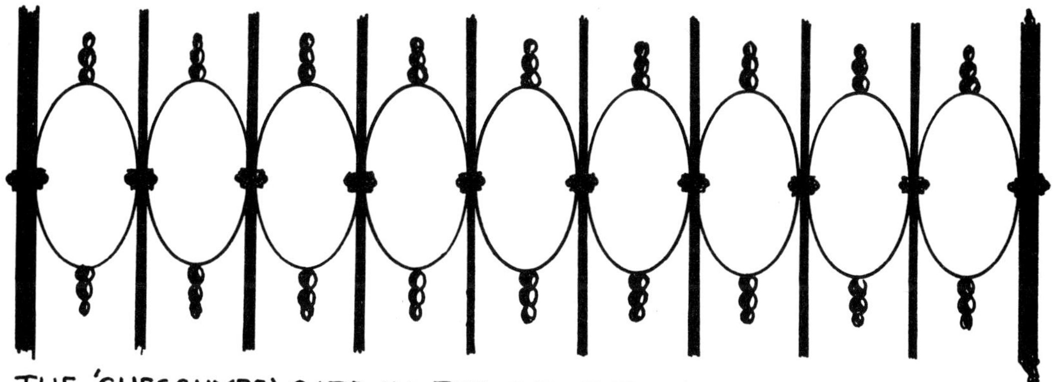

THE 'CHESSHYRE' GATE IN THE INDUSTRIAL MUSEUM, DERBY.

THE GARDEN GATE, THE HALL, TISSINGTON NR. ASHBOURNE, DERBYSHIRE.

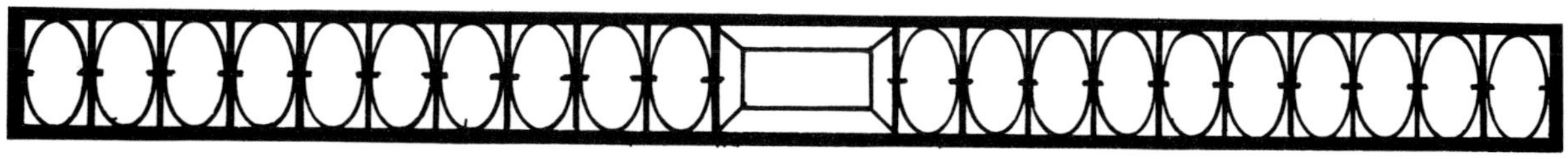

THE PARK GATES, OKEOVER, STAFFORDSHIRE. THE ABOVE IS A DIAGRAM OF THE 'STRETCHER' FRAME, FORMING THE BASE OF THE 'OVERTHROW'. THE ELLIPSES ARE, THEREFORE, WELL ABOVE EYE-LEVEL AND CONTRIBUTE TO THE 'UPWARD THRUST' WHICH CHARACTERISES THE WORK OF ROBERT BAKEWELL. THE SAME PLOY IS USED AT ALDENHAM PARK AND AT MELBOURNE ON THE ARBOUR.

The Bakewell Roundel....

Fig:15.

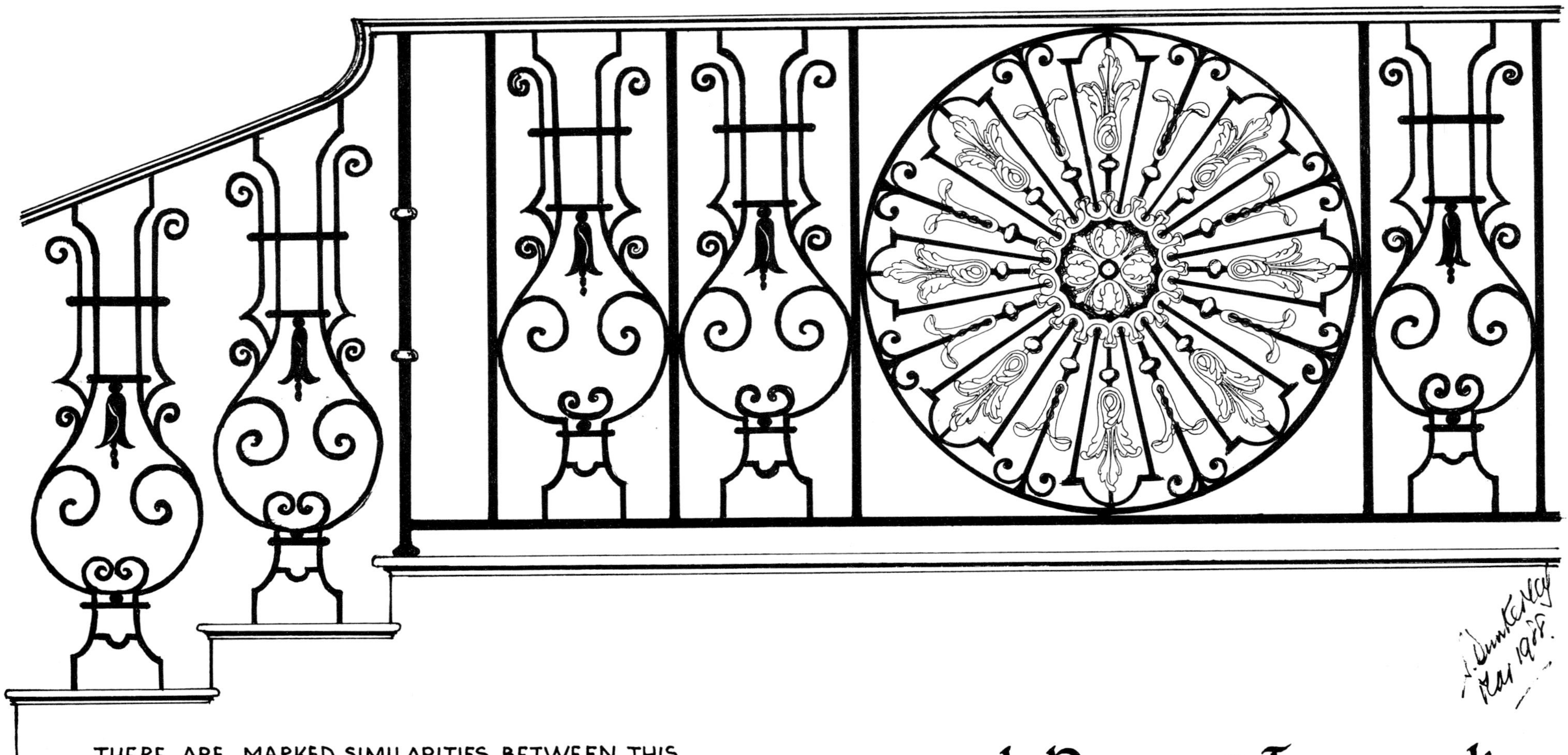

THERE ARE MARKED SIMILARITIES BETWEEN THIS ROUNDEL AND THOSE AT NEWNHAM PADDOCKS AND EATON HALL. SEE ALSO THE 'DARWIN' FANLIGHT IN DERBY'S INDUSTRIAL MUSEUM.

....and Perrow Ironwork

Bakewell motifs : Finials.

Fig. 16.

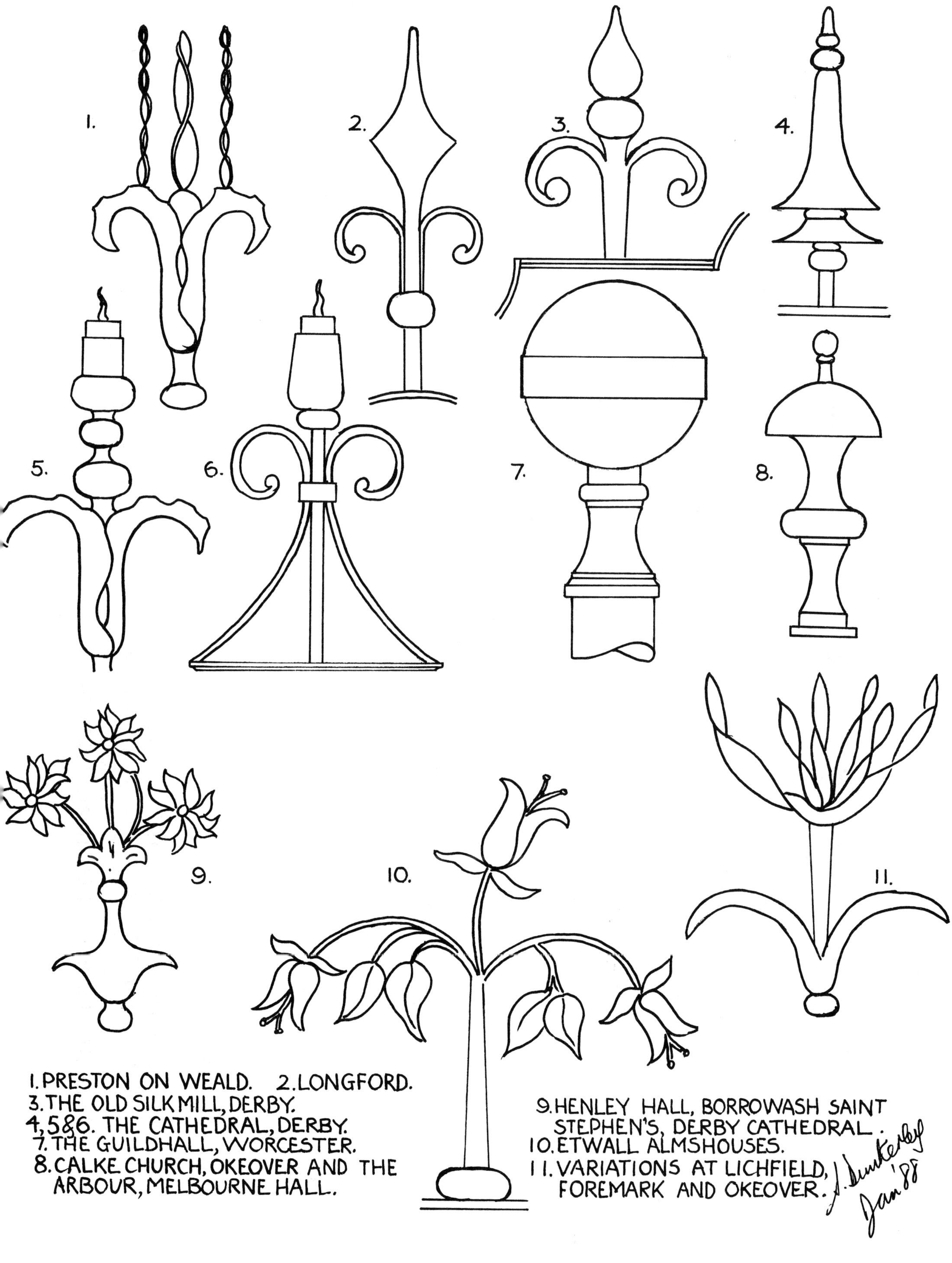

1. PRESTON ON WEALD. 2. LONGFORD.
3. THE OLD SILK MILL, DERBY.
4, 5 & 6. THE CATHEDRAL, DERBY.
7. THE GUILDHALL, WORCESTER.
8. CALKE CHURCH, OKEOVER AND THE ARBOUR, MELBOURNE HALL.
9. HENLEY HALL, BORROWASH SAINT STEPHEN'S, DERBY CATHEDRAL.
10. ETWALL ALMSHOUSES.
11. VARIATIONS AT LICHFIELD, FOREMARK AND OKEOVER.

A. Dunkerley Jan '88

Bakewell's Terminal Brackets

Fig: 17.

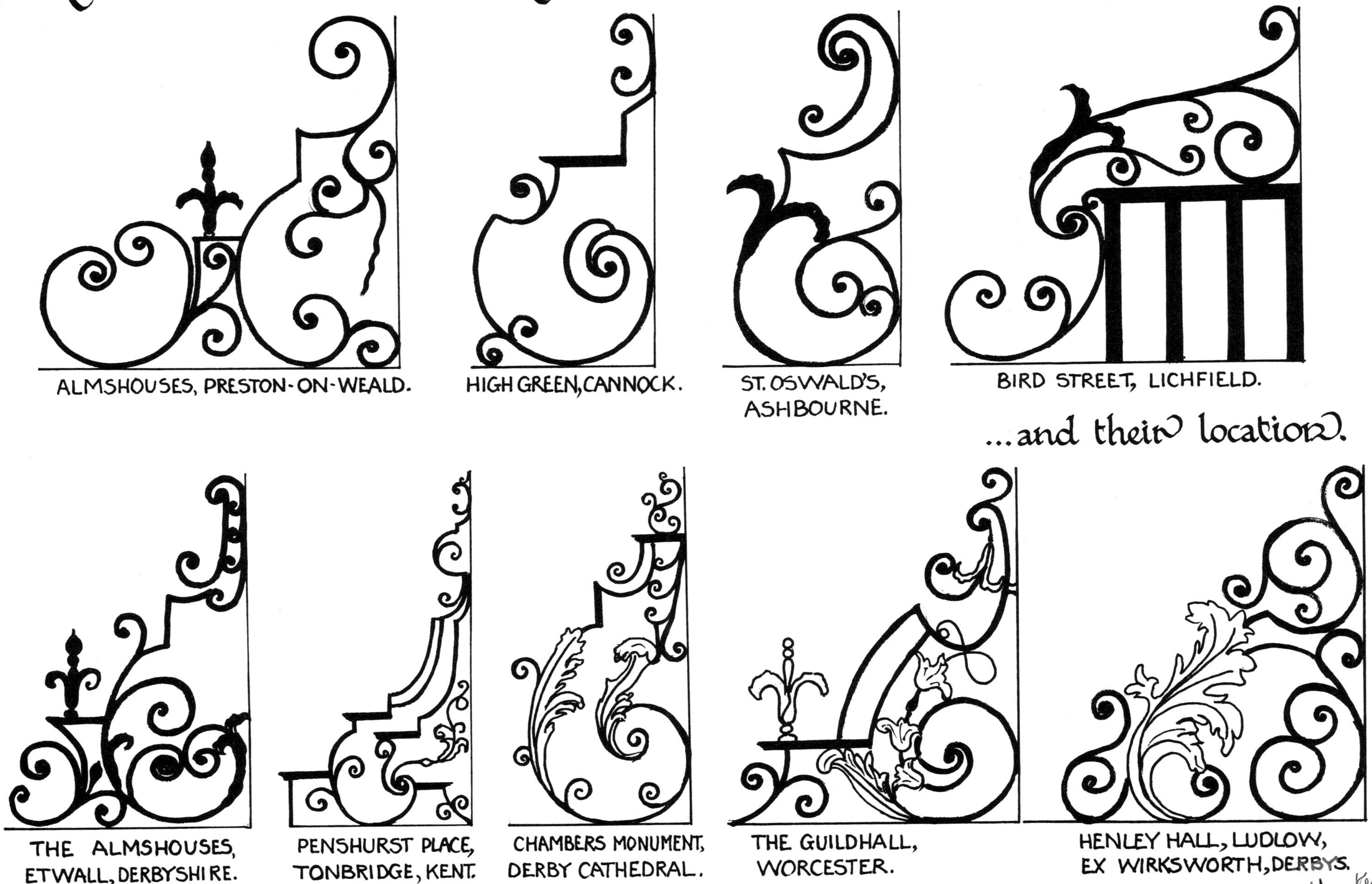

...and their location.

GAZETTEER

M denotes the RB mark; D work which is documented; A work attributed to Bakewell mainly on stylistic or circumstantial evidence.

CHESHIRE

Cholmondeley Castle. The home of the Marquess of Cholmondeley. Signposted from the A49 between Whitchurch and Tarporley. See the White screen, the Chapel screen and the Moat Garden balustrade. (D). Open to the public, gardens only, Sundays and Bank Holidays from Easter to September.

Dunham Massey. Owned by the National Trust. Near the town of Bowdon. 3 miles South West of Altrincham off the A56, Junction 19 off the M6 or Junction 7 off the M56. See the entrance grille in Bakewell style, the garden gate and the rails by the lake. (D). Open April to October every day except Friday.

THE CITY OF DERBY. Accessible from A516, A5111, A6, A61, A608, A52, A6005, A514 and A38.

The Cathedral Church of All Saints', Irongate.
See the Great Cavendish screen, the Altar base, the Communion rails, the Chambers' monument, the Corporation pew, the Consistory lamp bracket, and the Osborne gates at the West entrance. (D). Open daily during daylight.

The City Museum, The Strand.
See the 'blow-up' of the Acton gates (D) and the frame of the Bull's Head Inn sign. (A). Open each day except Sunday, Monday, Christmas Day, Boxing Day and Good Friday.

The Derby Industrial Museum, The Silk Mill, Full Street.
See the Chesshyre gate (A), the 'blow-up' of the Eaton Hall gate (A), The Fanlight from Darwin's House (D), the Rivett roundel (cypher) (D), the original Silk Mill roundel (also cypher) (M), a piece of tramsom bearing the RB Mark (M), and the beautiful Silk Mill gates. (M). There is much more of interest in this Museum — Rolls Royce engines, Cast ironwork from a local foundry, Clay and Stoneware. Open daily except Sunday, Monday, Christmas Day, Boxing Day and Good Friday.

Saint Werburgh's Parish Church, the Wardwick. Now redundant and for sale.
See the Font Cover and the Pelican Lectern. (D). Open Sunday mornings only.

DERBYSHIRE

Alvaston. Part of greater Derby. Lies at the junction of the A6 with the A5111, Derby to Castle Donington Road. Visit the Parish Church of Saint Michael, see the small but attractive Reredos. (A).

Ashbourne. A52 ex Derby; A517 ex Belper; A515 ex Sudbury; A515 ex Buxton; A523 ex Macclesfield and Leek. Saint Oswald's Parish Church, East entrance gates to Churchyard. (A).

Barton Hall, Church Broughton. Perron ironwork with balustrades. (D). This is a Private residence.

Borrowash. A52 Derby to Nottingham. Saint Stephen's Parish Church, Wrought iron screen, formerly at Hopewell Hall, Ockbrook. (A).

Calke Abbey, Ticknall. A514 Swadlincote to Melbourne. Owned by the National Trust. Due after serious and costly restoration to open to the public in 1989. In the Church at Calke the 'Plain but good' guard rails to the Harpur-Crewe tomb. On the Stable Cupola the Weather Vane. (D).

Duffield. A6 Derby to Belper. Saint Alkmund's Parish Church, Weather Cock. (A).

Etwall. A516 Derby to Hatton. Almshouses by the Church, Carriage gates. (A).

Foremark. Repton to Ingleby Road. Turn off to copse partly hiding the Church of Saint Saviour's. See the Communion rails (D) and behind the East end of the Church the Avenue gates and rails. (A).

Longford. The A515 Tutbury to Ashbourne road, turn off for Great Cubley, pass through Alkmonton to Longford. See the Carriage gates. (A).

Melbourne. Off A514 Derby to Swadlincote. Melbourne Hall. See the famous garden arbour - now a Grade 1 Listed building and the garden balustrade. (D). Open Easter to October, weekends only; mid June to mid September daily. (Closed Monday and Friday; open Bank Holiday Monday).

Radbourne. This hamlet lies between Mickleover and Kirk Langley. (A516 South West of Derby and A52 North East of Derby). Saint Andrew's Church, Weather Vane. (A).

Spondon. A52 South East of Derby. The Homestead, perron ironwork, well cared for. Owned by British Celanese from whom permission should be obtained. (A).

Tissington. Off the A515 North of Ashbourne. The Hall, Garden gate. (A).

KENT

Penshurst Place near Tonbridge. On B2716 off A26 Tonbridge to Tunbridge Wells or off B2027 Marlpit Hill to Tonbridge. See the Carriage gates, (M), formerly at Wingerworth, Derbyshire and a Royal residence in London. Open to the public every afternoon except Mondays but open on Bank Holidays.

LANCASHIRE

Burnley. A646 North West from Todmorden to Burnley. Towneley Hall Art Gallery. See the Staircase balustrade. (A). Open Monday-Friday 10-5. Sunday 12-5. Closed Saturday.

Manchester. The Church of Saint Anne, Saint Anne's Square.
See the Communion Rails. (A). The Church is beautiful and is open most days.

LEICESTERSHIRE

Ravenstone. Off the A50 Ashby-de-la-Zouche to Coalville Road. Saint Michael's Parish Church. See the most striking Communion rails. (A). (The Church is locked; enquire at the Vicarage).

Staunton Harold. Off the B587 Ashby to Melbourne road. Visit the Church of the Holy Trinity which is administered by the National Trust. See the Screen. (D). The Hall is now headquarters of the Ryder-Cheshire Mission. See the Staircase if you are able to obtain admission. It is of the same pattern as that at Okeover Hall and the Maister's House, Hull. (A). Open April-October, Wednesday-Sunday and Bank Holiday Mondays.

LONDON

Saint James's Place. Guard Rail Ironwork at the former residence of the Rt. Hon. Thomas Coke. (D).

UNITED STATES OF AMERICA

Long Island, New York. Visit the Old Westbury Gardens and see the Carriage gates (M) ex Combermere Abbey.

OXFORD

The Radcliffe Camera. This is a building based on a 'double octagon'. On its periphery are seven grilles in blue and gold - well worth the viewing. (D).

SHROPSHIRE

Henley near Ludlow. Off the B4346 Bridgnorth to Ludlow road, two miles North East of Ludlow. Henley Hall Gates and rails formerly at Wirksworth Hall, Derbyshire. (D).

Morville. Four miles North West of Bridgnorth on the A458. Aldenham Park gates and rails. (D).

Preston on Weald. Off the A518 Newport to Wellington road. Preston Almshouses, Gates and Pilasters. (The rails were cast at Coalbrookdale). (M).

Shrewsbury. The Abbey House Gates and rails. (A).

STAFFORDSHIRE

Cannock. High Green, former Council Offices. Gates and Pilasters. (A).

Chillington Hall. Off the M54 at Junction 2 or 3. Codsall Wood, South of Brewood. The home of Mr. and Mrs. Peter Giffard. See the Bowling Green Arch, gates and panels. Recently restored by Richard Quinnell of Leatherhead. (A). Open to the public on Wednesday afternoons during the summer months.

Lichfield. A38. Georgian House on Bird Street. Grade 2 listed building and so is the ironwork. (A).

Okeover. Leaving Ashbourne North on the A515 fork left for Thorpe, Dovedale. On the opposite side of the River Dove is the village of Ilam. Close by is Okeover Hall. The gates by the roadside are by Bakewell as are also others in the grounds. (D). This is a private residence.

Church Leigh near Uttoxeter. A522 South of Fole or A50 Burton to Stoke, Park Farm, typical Bakewell Gates. (A). Private residence.

WALES

Mold. A541 Trefnant to Wrexham or A494 Ruthin to Queensferry. **Leeswood.** The White Screen. Bakewell and the Brothers Davies are said to have worked together on this project. Bakewell most certainly was connected with the design. (A).

Ruthin. A494 from the A5 to Queensferry. **Llanhrydd.** Gates and panels recently re-assembled in a not too satisfactory manner. They had been badly damaged by a dust cart. (A).

WARWICKSHIRE

Newnham Paddox, near Monks Kirby. Off the B4112 Bedworth to Lutterworth road. Turn off at Withybrook. Permission is required as the gates are approached by a private road. A marvellous screen, the original parts are said to have been made by the Davies Brothers and the Wicket gates by Bakewell. (A).

WORCESTER. The Guildhall gates and rails - magnificent. (A).

YORKSHIRE

Hull. The Maister's House. Staircase of the same pattern as at Okeover and Staunton Harold. (D).

Wentworth Woodhouse. Northern College, Wentworth Castle, Stainborough. Between Rotherham and Barnsley. See the perron ironwork. (D). View by arrangement with the College authorities.

York. Go to Beningbrough Hall. See the four grilles in the main hall. (A). National Trust. Open daily (except Monday and Friday) May-October 12-5. At Easter 1989 open Good Friday through to following Thursday.

As opening times may be altered subsequent to the compilation of the Gazetteer it is advisable to check when planning a visit.

ACKNOWLEDGEMENTS

The Author expresses his sincere thanks to the following for permission to publish photographs, quote material and for supplying information.

The Viscount de Lisle V.C., K.G. Penshurst Place, Kent.
Group-Captain J. R. Bradshaw (Retd.) Penshurst Place.
The Most Honourable the Marquess of Cholmondeley, Cholmondeley Castle, Cheshire.
Mrs. Jean Grimes, Secretary, the Estate Office, Cholmondeley Castle.
Lord Ralph Kerr, Melbourne Hall, Derbyshire.
Mrs. G. A. Weston, Curator, Melbourne Hall.
Mr. E. J. Saunders, R.I.B.A., of Ticknall, Keeper of Muniments, Melbourne Hall.
Sir Peter Walker-Okeover, Okeover Hall, Ashbourne, Derbyshire.
Mr. Peter Giffard, Chillington Hall, Wolverhampton.
Mr. and Mrs. J. W. Pratt, Barton Hall, Derbyshire.
Mr. Henry Harpur Crewe.
John Bowen Evans of Burton-upon-Trent.
Mr. E. J. Knobbs, Park Hall Farm, Staffordshire.
Mr. D. W. H. Nielson, Catton Hall, Burton-upon-Trent.
Mr. J. M. Hurt, 111 Old Westbury Gardens, New York.
Miss Laura J. Carpenter, Archivist, Westbury Gardens.
Mr. I. A. Trelter, Schloss Charlottenburgh, West Berlin.
Mrs. Maureen Jones, Llanrhydd Hall, Ruthin, Clwyd.
Miss Rosemary Meynell.
Mr. Peter King, British Artist Blacksmiths Association.
Mr. Peter Walker, Chapel Cottage Forge, Anslow, Burton-upon-Trent, for checking the terminology of the line drawings and supplying information on restoration.
Mr. Michael Malleson.
Mr. Richard Quinnell for information on restoration.
The National Trust and Mr. P. E. Heath.
The Trustees of the Radcliffe Camera, Oxford.
Mr. I. F. Guest, Lincoln's Inn Fields.
Dr. Bruce Barker Benfield, Dept. of Western Manuscripts, Bodleian Library.
The Syndics of Cambridge University Library and Mr. A. E. B. Owen.
The British Library and Dr. Frances Harris.
The Wallace Collection and Miss S. Newell.
The Victoria and Albert Museum and Marian Campbell, Assistant Keeper, Metalwork.
Derby Museums and Mr. Brian Waters, Senior Keeper Industry and Technology.
Maxwell Craven, Keeper of Antiquities, Derby Museums.
Derbyshire County Library Service.
The Staff of the Local Studies Library, Derby.
Cheshire County Record Office and Mr. F. I. Dunn.
Derbyshire County Record Office and Miss Joan Sinar.
Staffordshire County Record Office: Mr. D. Fowkes and Mr. D. A. Stokes.

BIBLIOGRAPHY

BOOKS

M. Ayrton and A. Silcock. Wrought Iron and its Decorative Use. Country Life. 1929.
H. Batsford. The Cathedrals of England. Batsford. 1960.
G. Beard. Georgian Craftsmen and their Work. Country Life.
John Betjeman. Guide to the Parish Churches of England and Wales. Collins. 1980.
Marian Campbell. Introduction to Ironwork: the Victoria and Albert Museum. H.M.S.O. 1986.
Howard M. Colvin. Calke Abbey: A Hidden House Revealed. National Trust. 1985.
M. Craven and M. Stanley. The Derbyshire Country House.
Derbyshire Museum Service. Vol. I 1982. Vol. II 1984.
M. Craven. The Derby Town House. Breedon Books. 1987.
Ifor Edwards. Davies Brothers: Gatesmiths. Welsh Arts Council. 1977.
J. S. Gardner. English Ironwork of the 17th and 18th Centuries. Victoria and Albert Museum. 1978.
S. G. Gillam. The Building Accounts of the Radcliffe Camera. Clarendon Press. 1958.
G. J. Hollister-Short. Discovering Wrought Ironwork. 1970.
R. Lister. The Craftsman in Metal. Bell. 1966.
R. Lister. Decorative Wrought Ironwork in Great Britain. David and Charles. 1970.
Arthur Mee. The King's England. Hodder and Stoughton. 1937.
N. Pevsner. The Buildings of England. Penguin. 1978.
S. Sitwell. British Architects and Craftsmen 1600-1830. Century Classics. 1986.
Marcus Whiffen. Stuart and Georgian Churches outside London. Batsford. 1947.

PAMPHLETS

A Pilgrim's Guide to Derby Cathedral. R.S.M. O'Ferrall.
National Trust Pamphlet on Staunton Harold, available at the Church.

ARTICLES

In Country Life
Ifor Edwards. A Great Gatesmith re-assessed. September 27 1962.
Ifor Edwards. British Iron in a New York Garden. March 1980.
Gervase Jackson-Stopps. Cholmondeley Castle. July 19 and 26 1973.
Edward Saunders. Wrought Iron of a Master Smith. January 23 1980.
Arthur Oswald. Okeover Hall, Staffordshire. January 23 and 30 and March 19 1964.

In Derbyshire Life and Countryside.
Maxwell Craven. Vanished Town Houses. December 1963.
E. J. Saunders. Robert Bakewell of Derby. April 1982.

PHOTOGRAPHIC SOURCES

Plate 5.	The four Grilles of Beningbrough. The National Trust. Transparencies.
Plate 10.	The Towneley Hall Staircase. Towneley Hall Museum, Burnley. Transparency.
Plate 14.	The Table Frame. E. J. Saunders R.I.B.A.
Plate 18.	Penshurst Place. Group Captain Bradshaw.
Plate 18.	John Bowen Evans, Burton-upon-Trent. Mrs. M. Jones, Llanrhydd, Ruthin.
Plate 25.	Dunham Massey. Transparencies. The National Trust.
Plate 27.	Preston-on-Weald, John Bowen Evans.
Plate 29.	Leeswood Hall, John Bowen Evans.
Frontispiece	Old Westbury Gardens, Long Island, New York, photographed by Richard Cheek, Belmont, Mass.
Page 15.	Le Moulinet by Lancret. Schloss und Garten, West Berlin.
Page 20.	Staunton Harold Staircase. © *Country Life*.
Page 66.	Wentworth Woodhouse, West Front. © *Country Life*.

INDEX

SOME OTHER SCARTHIN BOOKS

Journey from Darkness. Gordon Ottewell. £2.40.
An adventure story for 9-12 year olds, following the hazardous journey of two pit-boys and a pony through the Derbyshire countryside of Victorian days.

Our Village: Alison Uttley's Cromford. Alison Uttley. £2.85.
Published for the Alison Uttley Centenary in 1984 and reprinted with an introduction by Denis Judd in 1987. A selection of her essays describe the Country Child's girlhood in Victorian Cromford. Illustrations by C. F. Tunnicliffe from the original editions.

First Loves: Life Stories of Victorian Dolls. Lilian McCrea. £6.95.
Large format hardback, with colour photographs of the dolls themselves from the author's collection. Lilian McCrea retells the stories, amusing and sad, which she heard from the original owners.

HISTORY TITLES

Hanged for a Sheep: Crime in Bygone Derbyshire. E. G. Power 1981. Second edition 1982. Reprinted 1988. £2.85.
Photographs and facsimile documents. A factual but entertaining survey of crime, law enforcement and punishment c. 1750-1850.

Ancient Wells and Springs of Derbyshire. Peter J. Naylor 1983. £2.25.
Photographs and diagrams. The only book on the natural waters of Derbyshire; with gazetteer of sites.

The Coal Mines of Buxton. Alan Roberts and John Leach. 1985. £3.75.
Photographs, maps and diagrams. An absorbing study of the forgotten pits and pit-men of the Buxton area.

St. John's Chapel, Belper: The Life of a Church and a Community. E. G. Power 1985. £1.95.
Fully illustrated. A history of the "Foresters' Chapel" and the people it served, from the thirteenth century to the present day.

Pauper's Venture: Children's Fortune. The Lead Mines and Miners of Brassington. Ron Slack 1986. £3.00.
A study of the lead-mining community of Brassington. Illustrated; with gazetteer of sites.

Joseph Whitworth: Toolmaker. Terence Kilburn 1987. £3.00.
An illustrated account of the great Victorian engineer.

"See that you hold fast to that which we leave you, Yea and teach your children its value, that never in the coming centuries their hearts may fail them or their hands grow weak."

Sir Francis Drake.

NOTES AND OBSERVATIONS

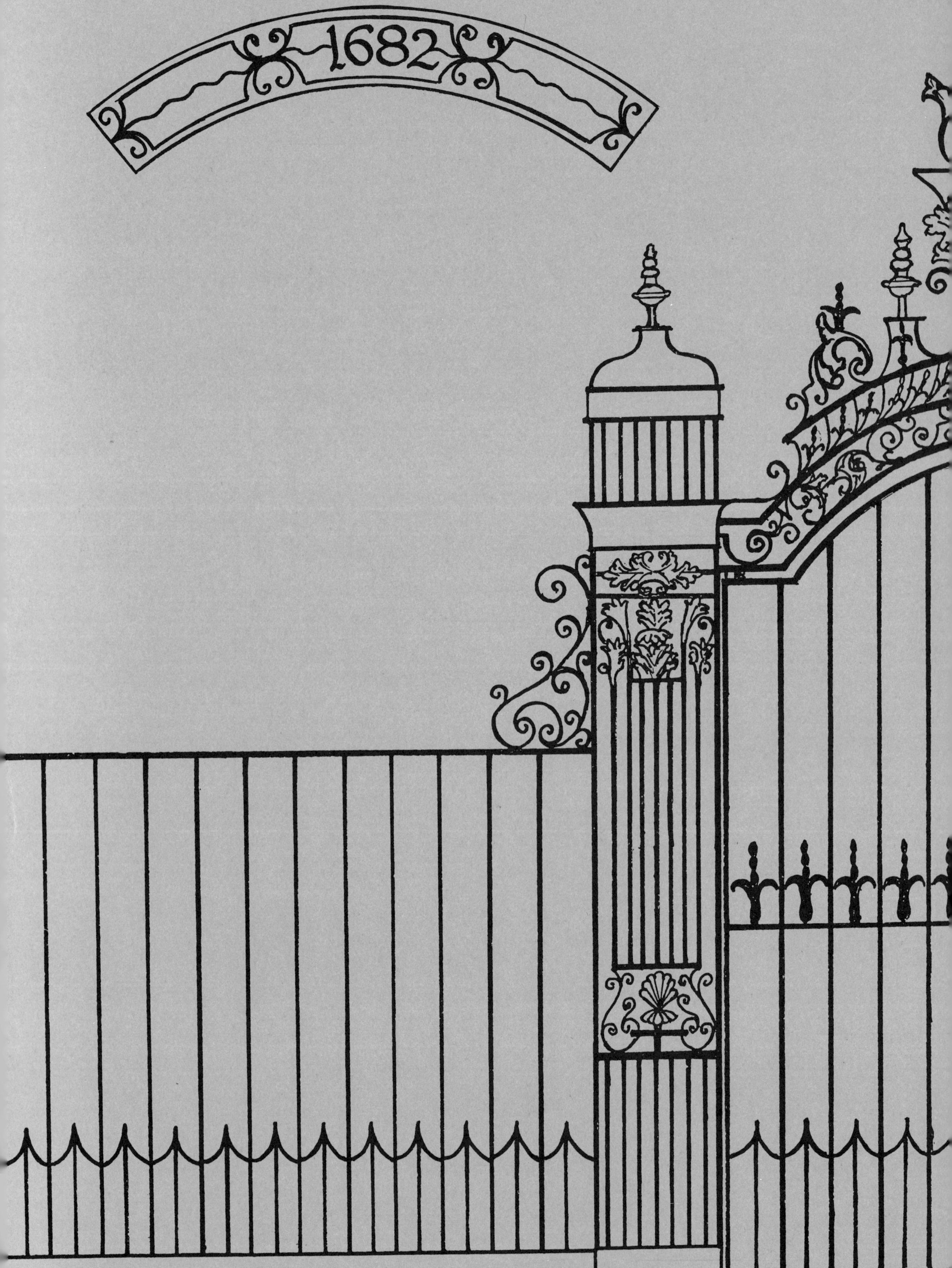
1682